Jonas Dimroth

Asymmetric Transfer Hydrogenation with Immobilized Catalysts

Jonas Dimroth

Asymmetric Transfer Hydrogenation with Immobilized Catalysts

Reusable catalysts for the production of chiral compounds

Südwestdeutscher Verlag für Hochschulschriften

Impressum/Imprint (nur für Deutschland/only for Germany)
Bibliografische Information der Deutschen Nationalbibliothek: Die Deutsche Nationalbibliothek verzeichnet diese Publikation in der Deutschen Nationalbibliografie; detaillierte bibliografische Daten sind im Internet über http://dnb.d-nb.de abrufbar.
Alle in diesem Buch genannten Marken und Produktnamen unterliegen warenzeichen-, marken- oder patentrechtlichem Schutz bzw. sind Warenzeichen oder eingetragene Warenzeichen der jeweiligen Inhaber. Die Wiedergabe von Marken, Produktnamen, Gebrauchsnamen, Handelsnamen, Warenbezeichnungen u.s.w. in diesem Werk berechtigt auch ohne besondere Kennzeichnung nicht zu der Annahme, dass solche Namen im Sinne der Warenzeichen- und Markenschutzgesetzgebung als frei zu betrachten wären und daher von jedermann benutzt werden dürften.

Verlag: Südwestdeutscher Verlag für Hochschulschriften GmbH & Co. KG
Heinrich-Böcking-Str. 6-8, 66121 Saarbrücken, Deutschland
Telefon +49 681 37 20 271-1, Telefax +49 681 37 20 271-0
Email: info@svh-verlag.de

Approved by: Berlin, TU, Diss., 2011

Herstellung in Deutschland:
Schaltungsdienst Lange o.H.G., Berlin
Books on Demand GmbH, Norderstedt
Reha GmbH, Saarbrücken
Amazon Distribution GmbH, Leipzig
ISBN: 978-3-8381-3039-2

Imprint (only for USA, GB)
Bibliographic information published by the Deutsche Nationalbibliothek: The Deutsche Nationalbibliothek lists this publication in the Deutsche Nationalbibliografie; detailed bibliographic data are available in the Internet at http://dnb.d-nb.de.
Any brand names and product names mentioned in this book are subject to trademark, brand or patent protection and are trademarks or registered trademarks of their respective holders. The use of brand names, product names, common names, trade names, product descriptions etc. even without a particular marking in this works is in no way to be construed to mean that such names may be regarded as unrestricted in respect of trademark and brand protection legislation and could thus be used by anyone.

Publisher: Südwestdeutscher Verlag für Hochschulschriften GmbH & Co. KG
Heinrich-Böcking-Str. 6-8, 66121 Saarbrücken, Germany
Phone +49 681 37 20 271-1, Fax +49 681 37 20 271-0
Email: info@svh-verlag.de

Printed in the U.S.A.
Printed in the U.K. by (see last page)
ISBN: 978-3-8381-3039-2

Copyright © 2012 by the author and Südwestdeutscher Verlag für Hochschulschriften GmbH & Co. KG and licensors
All rights reserved. Saarbrücken 2012

Acknowledgements

Many people helped me directly or indirectly in completing this work. I am grateful and indebted to all of them. First of all, I would like to thank Prof. Dr. Schomäcker for the supervision of this thesis, his great support and commitment, and for the many stimulating discussions we have had. I am also grateful to Prof. Dr. Haag for acting as a referee and for his encouraging and valuable support during our collaboration.

I am most deeply indebted to Dr. Uwe Schedler for giving me this opportunity and for his initiative, participation, and belief in this project. This thesis would not have been possible without him. I would also like to thank the whole team from PolyAn GmbH, especially Dr. Thomas Thiele for his permanent support and encouragement as well as Dr. Heike Matuschewski, Markus Heinrich, and Dr. Christian Heise for their help on various occasions. For his patience, the expert scientific support, and our helpful discussions I am very thankful to Prof. Dr. Wessig. I also wish to acknowledge the inspiration from Prof. Dr. Wandrey, who called my attention to ATH reactions.

Sincere thanks are given to Dr. Juliane Keilitz for the pleasant, motivating, and very fruitful collaboration as well as to all members of the TC 8 research group for the congenial atmosphere and for the help I received so many times! For assistance with measurements, the introduction into the handling of analytical equipment, and other kinds of help, I would like to give special thanks to Astrid Müller-Klauke, Anke Rost, Benjamin Beck, Christa Löhr, Gabi Vetter, Johnny Nachtigall, Dr. Kathrin Schneider, Le Anh Thu Nguyen, Dr. Maria Schlangen, Dr. Michael Schwarze, Dr. Oliver Schwarz, Riny Yolandha Parapat, Dr. Robert Frau, Dr. Sebastian Arndt, Dr. Silvia Czapla, Thorsten Otremba, and Xiao Xie. The assistance of Steffen Schrettl, Ronny Scherer, Frank Liebau, Claudia Jung, and Michael Stolarski is also gratefully acknowledged.

Furthermore, I would like to express my gratitude to the Berlin Senate State Department of Science, Research, and Culture for providing an Elsa-Neumann-Scholarship.

I thank my family for supporting me in all my endeavors. And I thank Jae-Yun Lee for being the best companion in life.

Abstract

Asymmetric transfer hydrogenation (ATH) is an operationally simple, safe, and environmentally benign catalytic method for the generation of chiral compounds. The catalysts used are generally transition metal complexes based on chiral ligands and either rhodium, ruthenium, or iridium. The high purity requirements on chiral substances, especially when they are used as intermediates in the production of pharmaceuticals, as well as the high catalyst costs make a complete catalyst separation essential and the reuse of the catalysts desirable. Thus, in order to increase the overall efficiency of ATH processes, in this thesis quantitatively separable and highly reusable catalysts were established via immobilization. New strategies were developed to synthesize linker-containing rhodium and ruthenium complexes, which could then be attached to surface-functionalized polymer supports. Functional heterogeneous ATH catalysts were generated from a modified version of a rhodium(III)-*p*-toluenesulfonyl-1,2-diphenylethylenediamine complex with tethered cyclopentadienyl unit immobilized on polymer chips and polymer beads. The supported catalysts were applied in the asymmetric transfer hydrogenation of aryl ketones in an aqueous solution of sodium formate, and excellent enantioselectivity and reusability were achieved. In determining the optimal reaction conditions, the pH of the solution was found to play a particularly decisive role in determining the activity and reusability of the catalysts, and a significant improvement was achieved when the reaction was run in an acidic medium. The results of kinetic experiments indicated that a second-order model describes the enantioselective conversion of acetophenone to phenylethanol under both basic and acidic conditions. However, significantly different rates and activation parameters suggested different mechanisms; acid may accelerate the reaction by changing the mode of the proton transfer. The catalytic system proved very simple to use and robust, and through upscaling of the reaction it was shown that there is a high potential for technical application to the ecologically and economically rational production of enantioenriched building blocks.

Contents

1	**Introduction**	**17**
2	**Fundamentals and State of the Art**	**19**
2.1	Chirality and Chiral Technology	19
2.2	Industrial Requirements	20
	2.2.1 Catalyst Requirements	20
	2.2.2 Product Requirements	22
	2.2.3 Overall Process Requirements	22
2.3	Enantioselective Catalytic Hydrogenation of Ketones	24
	2.3.1 Asymmetric Hydrogenation	24
	2.3.2 Asymmetric Hydroboration	24
	2.3.3 Asymmetric Hydrosilylation	25
	2.3.4 Asymmetric Biocatalytic Reduction	26
2.4	Asymmetric Transfer Hydrogenation	26
	2.4.1 Historical Background	26
	2.4.2 Hydrogen Donor Systems	27
	2.4.3 Catalysts	28
	2.4.4 Mechanistic Aspects	30
	2.4.5 Applications and Industrial Impact	34
	2.4.6 Immobilization	35
3	**Support Materials Applied**	**40**
3.1	Material Choice and Preparation	40
	3.1.1 Molecular Surface Engineering	41
	3.1.2 Material Properties	41
4	**Modification of a Ruthenium Catalyst**	**44**
4.1	Background	44
	4.1.1 Homogeneous Applications	44
	4.1.2 Immobilization Approaches	45
4.2	Initial Studies	46
	4.2.1 Preparation and Use of Diphosphine/Diamine-based Catalysts	46
	4.2.2 Strategies for Modification	47

Contents

- 4.3 Preparation and Use of the Modified PNNP 48
  - 4.3.1 Synthetic Pathway 48
  - 4.3.2 Attempts at Immobilization 51
  - 4.3.3 Concluding Remarks 52

5 Immobilization of a Rhodium Catalyst   54
- 5.1 Synthetic Modification 54
  - 5.1.1 Ligand Preparation 54
  - 5.1.2 Catalyst Formation 55
- 5.2 Catalytic Testing 58
  - 5.2.1 General Remarks 58
  - 5.2.2 Initial Experiments 59
  - 5.2.3 Effect of Temperature and Atmosphere 63
  - 5.2.4 pH Dependency of Activity, Enantioselectivity, and Reusability 64
  - 5.2.5 Effect of the Concentrations of Acetophenone and Sodium Formate 67
- 5.3 Kinetic and Mechanistic Investigations 68
  - 5.3.1 Basic Reaction Conditions 69
  - 5.3.2 Acidic Reaction Conditions 70
  - 5.3.3 Mechanistic Considerations 71
- 5.4 Practical Aspects 76
  - 5.4.1 Application-oriented Experiments 77
  - 5.4.2 Discussion on the Efficiency 78

6 General Conclusion and Outlook   81

7 Experimental   84
- 7.1 Equipment 84
  - 7.1.1 Instrumentation 84
  - 7.1.2 Laboratory Equipment 85
- 7.2 Synthesis Procedures 86
  - 7.2.1 Synthesis of Ru-PNNP 86
  - 7.2.2 Synthesis of tethered Rh(III)-catalyst 95
- 7.3 Characterization of Supports and Coupling 96
  - 7.3.1 Determination of Amine Loading 96
  - 7.3.2 Immobilization and Determination of Catalyst Loading 97
- 7.4 Catalysis Experiments 98
  - 7.4.1 Homogeneous Catalysts 98
  - 7.4.2 Heterogenized Catalysts 98

Bibliography   100

Appendix 110

List of Figures

2.1	The BINAP ligand	25
2.2	Ligands successfully used in ATH operations	29
2.3	Intermediates produced via ATH in scale-up studies	35
3.1	Support materials applied	40
3.2	Schematic illustration of the support materials	42
3.3	Loading with amino groups vs. loading with rhodium	42
4.1	Diphosphine/diamine ligand and ruthenium complex	44
4.2	Immobilized ruthenium-PNNP complexes	45
5.1	Wills´s tethered Rh-TsDPEN catalyst	54
5.2	Supported Catalysts	57
5.3	IR spectra as proof of coupling	57
5.4	Use of supported catalysts	59
5.5	Reproducibility of the standard experiment using the PB-supported catalyst	61
5.6	Variation of temperature and atmosphere within the flask	64
5.7	Multiple catalyst reuse under different pH conditions	66
5.8	Impact of the concentrations of substrate and hydrogen donor	68
5.9	Influence of the concentration of acetophenone on the suspension of beads	68
5.10	ATH of acp under moderately basic conditions	69
5.11	Eyring plot for basic conditions	70
5.12	ATH of acp under acidic conditions	71
5.13	Eyring plot for acidic conditions	71
5.14	Ru-TsDPEN species under different conditions	72
5.15	Activity and enantioselectivity over pH	74
5.16	Alternative transition states under acidic conditions	75
5.17	Technically relevant tests	77

List of Schemes

2.1	The CBS reduction reaction	25
2.2	The MPV reduction and Oppenauer oxidation	27
2.3	Inner sphere mechanism	31
2.4	Concerted outer sphere mechanism	32
2.5	Proposed outer sphere mechanism in aqueous media	33
2.6	Covalent attachment of a TsDPEN derivative to polystyrene supports	37
2.7	Crosspolymerization of a TsDPEN-based ligand monomer	38
4.1	ATH of acetophenone as the benchmark reaction	45
4.2	Preparation of the unmodified PNNP ligand and ruthenium complex	46
4.3	Preparation of key intermediate 7	49
4.4	Preparation of key intermediate 12	49
4.5	Preparation of the modified Ru-PNNP complex	50
4.6	Coupling of modified Ru-PNNP	51
4.7	Coupling of modified PNNP	51
5.3	Introduction of methyl adipoyl chloride	55
5.1	Introduction of hex-5-ynoic acid-tert-butylester — first attempt	55
5.2	Introduction of hex-5-ynoic acid-tert-butylester — second attempt	55
5.4	Preparation of the supported modified tethered Rh-TsDPEN catalyst	56
5.5	ATH of aryl ketones in an aqueous medium	59
5.6	ATH of aryl ketones in formate/water	67
5.7	Proposed mechanism under different pH conditions	73

List of Tables

2.1	Principles of green chemistry and green engineering	23
2.2	Commonly used hydrogen donor systems	28
4.1	Performance of Ir-PNNP	47
4.2	Attempts at immobilization to aminated supports	52
5.1	Performance of different Rh-TsDPEN variants	60
5.3	Recycling of PC-supported catalyst with intermediate washing	61
5.2	Recycling of PC-supported catalyst without intermediate washing	62
5.4	Scope of tested substrates	62
5.5	ATH of acetophenone under different pH conditions	65

Abbreviations and Symbols

acp	acetophenone
AH	asymmetric hydrogenation
aq	*aqua*
ATH	asymmetric transfer hydrogenation
ATR	attenuated total reflection
BINAP	2,2'-bis(diarylphosphino)-1,1'-binaphthyl
Boc	*tert*-butyloxycarbonyl
cat	catalyst
CBS	Corey–Bakshi–Shibata
CDI	1,1'-carbonyldiimidazole
COD	cyclooctadiene
Cp	cyclopentadienyl
DAD	diode array detector
DCC	N,N'-dicyclohexylcarbodiimide
DCM	dichlormethane
DFT	density functional theory
DIC	N,N'-diisopropylcarbodiimide
DIPEA	diisopropylethylamine
DMAP	4-(dimethylamino)pyridine
DMF	dimethylformamid
DMSO	dimethyl sulfoxide
DPEN	1,2-diphenylethylenediamine
DSC	differential scanning calorimetry
ee	enantiomeric excess
EI	electron ionization
ESI	electron spray ionization
FID	flame ionization detector
Fmoc	9-fluorenylmethyloxycarbonyl
FT-IR	fourier transform infrared spectrophotometer
GC	gas chromatography
h	Plank constant [$6.626 \cdot 10^{-34}$ J s]
HOAt	1-hydroxy-7-azabenzotriazole

Abbreviations and Symbols

HOBt	hydroxybenzotriazole
HPLC	high performance liquid chromatography
HRMS	high-resolution mass spectrometry
ICP	inductively coupled plasma
IPA	2-propanol
IR	infrared
J	coupling constant [Hz]
k	rate constant
k_B	Boltzmann's constant [$1.381 \cdot 10^{-23}$ J K^{-1}]
MPV	Meerwein–Ponndorf–Verly
MS	mass spectrometry
n	amount of substance [mol]
NHS	N-hydroxysuccinimid
NMR	nuclear magnetic resonance
OES	optical emission spectroscopy
PB	polymer bead
PC	polymer chip
PDMS	polydimethylsiloxane
PE	polyethylene
PEG	polyethylene glycol
PNNP	N,N'-bis[o-(diphenylphosphino)-benzyl]cyclohexane-1,2-diamine
PP	polypropylene
ppm	parts per million
PPNCl	bis(triphenylphosphoranylidene) ammonium chloride
PS	polystyrene
R	universal gas constant [8.3145 J mol^{-1} K^{-1}]
rpm	revolutions per minute
rt	room temperature
S	selectivity
S/C	substrate to catalyst ratio
SDS	sodium dodecyl sulfate
STY	space time yield [g L^{-1} h^{-1}]
T	absolute temperature [K]
TBTU	o-(benzotriazol-1-yl)-N,N,N',N'-tetramethyluronium tetrafluoroborate
TEAF	triethylamine/formic acid
tert	tertiary
TFA	trifluoroacetic acid
TH	transfer hydrogenation
THF	tetrahydrofuran

TLC	thin layer chromatography
TOF	turnover frequency [h^{-1}]
TON	turnover number
TS	transition state
TsDACH	N-(p-toluenesulfonyl)-1,2-diaminocyclohexane
TsDPEN	N-(p-toluenesulfonyl)-1,2-diphenylethylenediamine
TTN	total turnover number
UV-Vis	ultraviolet-visible spectral region
Y	yield

Greek Characters

δ	chemical shift [ppm]
$\Delta G^\ddagger$	Gibb´s energy of activation [kJ mol^{-1}]
$\Delta H^\ddagger$	enthalpy of activation [kJ mol^{-1}]
$\Delta S^\ddagger$	entropy of activation [J mol^{-1} K^{-1}]
λ	wavelength [nm]
ν	wavenumber [cm^{-1}]
Φ	phenyl

1 Introduction

The growing demand for enantiopure chiral compounds, which play an important role in the fabrication of pharmaceuticals [1], is accompanied by increasing acknowledgement of the "need for more environmentally acceptable processes in the chemical industry" [2]. Maximal efficiency and minimal hazard are the criteria for the design of new processes to meet the complex needs of society and the economy. Generally, catalytic production is favorable over non-catalytic since catalysts are capable of making chemical reactions more energy-efficient; this is the reason why catalysis-based manufacturing accounts for about 85 % of all chemical production [3]. An additional benefit is that the use of asymmetric catalysts in the fabrication of chiral substances leads to the multiplication of the precious chiral information. For current and future asymmetric production processes, therefore, the optimization of catalytic methods is of great importance.

Of the several asymmetric catalytic methods suitable for the enantioselective reduction of prochiral substrates, the asymmetric transfer hydrogenation (ATH) of imines and ketones is considered particularly useful because the reaction proceeds under mild conditions without the need for sophisticated equipment and hazardous reactants. Furthermore, intensive investigation in the last three decades has afforded a broad knowledge about ATH reaction mechanisms, the use of hydrogen donors, and the development of highly active and enantioselective catalysts. These catalysts are usually homogeneously applied complexes made of chiral organic ligands and transition metals; this is a drawback since both components are expensive and many transition metals are toxic. Thus, for industrial applications a complete removal of the catalyst from the reaction mixture is required to guarantee metal-free products, and the reuse of the catalyst is highly desired to reduce costs and improve the environmental impact of the process.

Heterogeneous catalysis suggests an approach to addressing this problem. Heterogeneous catalysts are separated from the reactants by virtue of their physical state, thus facilitated product separation and multiple use of each catalytic site over an extended lifetime are possible. With regard to other important issues, however, homogeneous catalysts are favorable. They are generally more active and more selective, reaction conditions are usually milder, mass transfer and diffusion problems are less severe, modification of electronic and steric properties is possible, and mechanistic investigation is easier [4]. To combine the best features of both heterogeneous and homogeneous catalysis, researchers often use catalyst immobilization. Of the various methods of catalyst immobilization, heterogenization via covalent attachment might be the most often applied. Nevertheless, making a homogeneous catalyst heterogeneous usually requires an elaborate modification of the ligand as well as a suitable support and efficient coupling, and thus increases the catalyst cost contribution to the overall process. Additionally, in many cases activity of the the catalyst decreases upon immobilization. Hence, the potential advantages of an immobilized system will not be utilized as long as they do not compensate for the higher costs. Since industrial processes based on immobilized asymmetric

1 Introduction

catalysts remain scarce [5], it seems that this condition has not yet been achieved. Both academia and industry are therefore called on to search for new routes to efficient immobilized catalytic systems, and to provide a fundamental understanding of how these systems work.

It was thus the objective of this thesis to generate new heterogenized catalysts and elucidate their strengths and weaknesses with regard to a potential application in ATH processes on laboratory or industrial scales. The aim was to develop an easy to use and efficently recyclable catalytic system and to achieve a comprehensive understanding of how the reaction conditions determine catalyst performance. To efficiently achieve these aims, a joint project was established bringing together expertise from academia and industry. Major parts of the experimental work, in particular all activities related to surface-functionalization, were done at the laboratories of PolyAn GmbH, Berlin. Analytics and performance tests were carried out at the Institute of Chemistry at the Technische Universität Berlin. Substantial support was received from the Institute of Chemistry and Biochemistry of the Freie Universität Berlin.

This thesis begins with an overview of the basic theoretical aspects and the fundamental developments in the field of ATH, followed by a description of the approaches used and the results obtained. First, I briefly describe the support materials which were applied. The modification, coupling, and experimental application of the transition metal complexes are then discussed separately for each approach. Finally, the major results and conclusions are reconsidered in a general discussion. Experimental details are given at the end of the dissertation.

2 Fundamentals and State of the Art

2.1 Chirality and Chiral Technology

Chirality (from the ancient Greek word ἡ χείρ, hand) or handedness is a phenomenon of fundamental importance in biology and thus in all life science-related technologies. An object is denoted chiral when it has a non-superimposable mirror image, just like a human hand. A pair of mirror image molecules is called a pair of enantiomers (ἐναντίος, opposite), with one having an R-configuration and the other an S-configuration according to the Cahn-Ingold-Prelog rules. Usually, three forms of chirality are distinguished: central (point), axial, and planar chirality [6]. Most important in the context of this thesis is central chirality, which is present when four different residues are attached to a carbon atom.

Different enantiomers of a given molecule can exhibit dramatically different biological activities. For example, the S,S-form of ethambutol is used as a tuberculostatic whereas the R,R-form may cause blindness [7]. Although the relevance of optically pure drugs had been known for some time, the chiral switch — the substitution of racemic for enantiopure pharmacological compounds — did not begin until the 1990s when the U.S. Food and Drug Administration (FDA) decreed new rules for the development of stereoisomeric drugs [8].

The production of pure enantiomers is a challenging task, since in any (conventional) reaction of achiral starting materials to chiral products these products will be formed as racemates, i. e., 50:50 mixtures of both enantiomers [9]. Currently, chiral pool methods or racemic resolutions are the approaches most commonly applied for the production of chiral compounds [5, 10]. Thus, either enantiopure building blocks provided by nature are used as starting materials, or the enantiomers produced are separated via crystallization, kinetic resolution, or chromatography [11, 12]. The first approach, however, is limited due to a limited number of accessible compounds which are not necessarily suitable, so that many steps with attendant losses in yield may be required for the synthesis of the desired product. The latter approach, the formation and separation of racemic products, has the drawback that 50 % waste is generated.

A generally more atom-economic and potentially more cost-efficient method is thus the use of asymmetric catalysts, which act as multiplicators of their chiral information. The following section provides an overview of the requirements that those catalysts have to meet to be applicable to industrial processes. A list of (academically) successful examples of enantioselective (chemo)catalytic approaches includes hydrogenation, transfer hydrogenation, hydrosilylation, and hydroboration of unsaturated compounds, epoxidation of allylic alcohols, vicinal hydroxylation, hydrovinylation, hydroformylation, cyclopropanation, and isomerization of olefins, propylene polymerization, organometallic addition to aldehydes, allylic alkylation, organic halide-organometallic coupling, aldol type reactions, and Diels-Alder and ene reactions [13].

2.2 Industrial Requirements

Requirements for catalytic chemical fabrication include *overall process requirements*, *product requirements*, and *catalyst requirements*, all of which are interdependent to some extent and thus lead to different points of view for how to deal with analogous problems. Enantiopure products are often subjected to the special conditions of pharmaceutical fabrication as they are processed to medicinal products in large part [14]. These conditions include, of course, the general requirements in chemical production, but are usually somewhat broader and more stringent.

Unfortunately, the definitions of many terms and key figures (performance indicators) which are employed to describe and compare catalyst or process properties, vary among the fields in which they are used. Furthermore, there is no general agreement on the conditions under which those key figures are to be determined. Thus, a reliable comparison is only possible if the units are known and the reaction conditions are roughly equal. The following definitions represent the general understanding in the context of organometallic catalysis with a focus on asymmetric synthesis. A general overview is given, for instance, by Behr [15], Hagen [16], and Weitkamp and Gläser [17].

2.2.1 Catalyst Requirements

The applicability of a catalyst in an industrial process depends mainly on three properties:

- activity
- selectivity
- stability.

Although a general statement is difficult to make due to the diverse industrial applications with their individual requirements on the catalysts, according to Hagen "the target quantities should be given the following order of priority: selectivity > stability > activity" [18]. Catalyst testing is mostly performed with simplified benchmark substrates, but for example in the production of pharmaceuticals, highly functionalized starting materials are often employed. Thus, the substrate scope and the functional group tolerance are also of great importance [19]. Further important aspects include, of course, economic issues. Although normally more important in the fabrication of low-priced chemicals, the catalyst costs can become crucial in asymmetric production processes where chiral metal complexes consisting of expensive noble metals and often even more expensive chiral ligands are applied. The contribution of the catalyst to the total cost, however, has to be discussed in the context of the overall process (see below).

Selectivity The selectivity, S_P, of a given reaction with more than one possible product is calculated by the ratio of the amount of the desired product (P) to the amount of substance of the starting material (A) which has been converted, taking into account the stoichiometric coefficients of the reactants (Equation 2.2.1). S_P refers to only one specific product and starting material, as do the conversion (only starting material) and the yield.

$$S_P = \frac{n_P \, |\nu_A|}{(n_{A,0} - n_A) \, |\nu_P|} \tag{2.2.1}$$

The conversion, X_A, represents the ratio of converted starting material to employed starting material (Equation 2.2.2). The yield, Y_P, is defined as the ratio of the amount of the desired product actually formed to the theoretical maximum amount (Equation 2.2.3). It is thus the product of conversion and selectivity.

$$X_A = \frac{n_{A,0} - n_A}{n_{A,0}} \tag{2.2.2}$$

$$Y_P = \frac{n_P \, |\nu_A|}{n_{A,0} \, |\nu_P|} = \text{conversion} \cdot \text{selectivity} \tag{2.2.3}$$

There is, however, more than one selectivity in many processes. In the special case of asymmetric catalysis, where prochiral starting materials are converted to chiral products, the **enantio**selectivity is of decisive importance. It is quantified by the ratio of the excess amount of the major enantiomer, e.g., the R-form, to the sum of the amounts of both enantiomers (Equation 2.2.4). When no further purification is possible, the enantiomeric excess (ee) provided by the catalyst should be higher than 99 % for pharmaceuticals, but since this is rather seldom, lower ees (>90 %) are usually acceptable. For agricultural chemicals 80 % ee can be sufficient [14, 20].

$$ee \, [\%] = \frac{R - S}{R + S} \cdot 100 \tag{2.2.4}$$

Stability The stability is strongly related to the catalyst lifetime, which is limited due to metal leaching, poisoning, or other types of deterioration resulting from chemical, thermal, or mechanical stress. Although the turnover number (TON) does not represent a direct quantification of the stability or lifetime of a catalyst, it implicitly provides information thereof by indicating how productive the catalyst is under the given reaction conditions. The TON is defined as the ratio of the amount of converted starting material to the amount of catalyst employed. Thus, it specifies the number of catalytic cycles run by one catalyst site under defined conditions (Equation 2.2.5).

$$\text{TON} = \frac{\text{converted starting material [mol]}}{\text{amount of catalyst employed [mol]}} \tag{2.2.5}$$

For the cost-efficient fabrication of rather cheap mass-marketed products, a TON of more than 20 000 is usually required. The small-scale production of compounds with higher added value can be profitable at a TON of 1 000 [21]. In a batchwise process in which a certain amount of the catalyst is repeatedly used, the TON of each batch can be summed to the *total* turnover number (TTN, Equation 2.2.6), which indicates the total productivity of the reused catalyst.

$$\text{TTN} = \sum \text{TON} \tag{2.2.6}$$

Another indicator often used by researchers is the substrate to catalyst ratio (S/C), i.e., the ratio of the total amount of substrate employed to the amount of catalyst employed. Information about the catalyst´s productivity, however, is only provided when the conversion or yield is additionally quoted (which effectively gives the TON).

Activity The catalyst activity is in most cases quantified by the turnover frequency (TOF), which is defined as the number of molecular reactions or catalytic cycles per unit time (Equation 2.2.7). For instance in enantioselective hydrogenation processes, the TOF at a conversion of more than 95 % ought to be above 500 h^{-1} for small-scale and above 10 000 h^{-1} for large-scale productions [14, 20].

$$\text{TOF } [h^{-1}] = \frac{\text{converted starting material [mol]}}{\text{amount of catalyst employed [mol]} \cdot \text{time [h]}} = \frac{\text{TON}}{\text{time [h]}} \qquad (2.2.7)$$

2.2.2 Product Requirements

In addition to the requirements regarding the enantiomeric purity (>99 % ee for pharmaceuticals, >80 % ee for agrochemicals), the main problem from the point of view of the product is the contamination with other compounds. Usually, a product purity of more than 99 % with metal residues below 10 ppm is reqired in the production of pharmaceuticals [5]. Thus, in addition to a high selectivity, the efficent separation of the catalyst is of decisive importance in the production of fine chemicals and pharmaceuticals.

2.2.3 Overall Process Requirements

The benefit of an (asymmetric) catalytic step within the production of a target molecule should be judged in the context of the whole process. Today´s industrial fabrication processes have to meet the following requirements — which are by no means mutually exclusive [22]:

- cost efficiency
- safety
- environmental friendliness.

Cost Efficiency Ideally, a chemical fabrication process provides a high product yield at low cost [23]. Thus, yield relative to proportional costs is a viable way of determining the efficiency of a given reaction. A detailed discussion of economic aspects is outside the scope of this thesis; with regard to the catalyst, however, there is a rule of thumb saying that its cost contribution to the total cost price of the product should not be more than 5 % [24]. Generally, the lower the added value of the compounds produced the more important the catalyst´s performance and cost as well as process optimization. In contrast, the decisive factors for the development of compounds with very high added value are "the total synthetic features and time issues together with intellectual property aspects and risks of constraints" [25].

A useful key factor for the determination of the efficiency of a (catalytic) reaction is the space time yield (STY). It indirectly indicates a part of the investment costs, since it includes the volume of the reactor.

Table 2.1: Principles of green chemistry and green engineering.

	Green chemistry	Green engineering
1	Waste prevention	Inherent rather than circumstantial
2	Atom economy	Prevention instead of treatment
3	Less hazardous syntheses	Design for separation
4	Designing safer chemicals	Maximize efficiency
5	Safer solvents and auxiliaries	Output-pulled versus input-pushed
6	Design for energy efficiency	Conserve complexity
7	Use of renewable feedstocks	Durability rather than immortality
8	Reduce derivatives	Meet need, minimize excess
9	Use of catalysis	Minimize material diversity
10	Design for degradation	Integrate material and energy flows
11	Analysis for pollution prevention	Design for commercial "afterlife"
12	Inherently safer chemistry	Renewable rather than depleting

STY is most commonly defined as the amount (specified as mass) of the desired product, which is formed within a given time in a given reaction volume (Equation 2.2.8).

$$\text{STY } [\text{g L}^{-1} \text{h}^{-1}] = \frac{\text{desired product [g]}}{\text{total reaction volume [L]} \cdot \text{time [h]}} \quad (2.2.8)$$

Safety and Environmental Issues There is an increasing demand for and interest in the development of sustainable and inherently safe processes in the chemical industry. Anastas and co-workers formulated 12 principles of *green chemistry* [26] and *green engineering* [27], which have been widely recognized as a useful guideline for more favorable approaches in chemical fabrication and process design. Some of these principles are redundant in a way; nevertheless, they are worth mentioning at least in an abbreviated form (see Table 2.1). Major importance in the context of "green" production is given to the minimization of waste, that is, all matter produced which is not part of the product at the end of the chemical process. The degree of waste prevention is expressed by the *atom economy* or *atom efficiency* of a given reaction or process, which — in addition to yield and selectivity — has become "the third element of the triadic goal that any synthetic chemist should seek" [28]. Several key factors are used to make the atom economy a comparable value. Arguably the most prevalent is the E-factor (Equation 2.2.9), introduced by Sheldon in 1992 [29].

The optimal value of the E-factor is zero, since it represents the ratio of the mass of the waste generated to the mass of the desired product. Conceptual limitations of this key measure become evident when one considers that the definition of waste is anything at the end of the process that is not the desired product; unconverted starting materials, additives, side products, solvent losses, and even the mass of fuel corresponding to the required energy can be taken into account. Water, however, is not considered as waste; only the compounds contained in the water are counted. Thus, substances are either weighted with zero (water) or with one (all others), not taking into account the real potential for harm of each individual compound. Other authors have attempted to overcome these limitations by proposing a more sophisticated weighting of toxicity and the waste-associated hazards of different types of substances [28]. Nevertheless, when used in addition to other criteria, the E-factor may be a powerful instrument for the comparison of different pathways which lead to the same product. It seems particularly useful when the calculation is simplified to the ratio of the mass of raw materials minus the mass of the desired product to the mass of the desired product, thus excluding all waste streams which may be difficult to characterize [30].

$$\text{E-factor} = \frac{\text{mass of waste}}{\text{mass of desired product}} \quad (2.2.9)$$

2.3 Enantioselective Catalytic Hydrogenation of Ketones

In addition to asymmetric transfer hydrogenation, there are four important catalytic methods to enantioselectively reduce (unactivated) prochiral ketones to chiral alcohols: asymmetric hydrogenation, asymmetric hydroboration, asymmetric hydrosilylation, and asymmetric biocatalytic reduction. All of these methods will be briefly presented in the following, though asymmetric transfer hydrogenation is discussed in more detail in Section 2.4.

2.3.1 Asymmetric Hydrogenation

The asymmetric hydrogenation (AH) of ketones based on ruthenium(II)-diphosphine catalysts established proof of concept during the 1980s and is now an industrially established method [31]. The advancement of AH is inextricably linked to R. Noyori who was decorated for his contributions to the field with the Nobel Prize in chemistry in 2001 [32]. Noyori and co-workers developed the 2,2'-bis(diphenylphosphino)-1,1'-binaphthyl (BINAP) ligand (Figure 2.1), which is—together with its derivatives—the basis for the most successful catalysts for AH operations [33, 34]. However, ruthenium(II)-BINAP complexes were initially only used as catalysts for the reduction of β-keto esters [35], until further progress extended the applicability to the hydrogenation of simple unactivated ketones such as acetophenone. It was found that the activity and productivity of AH reactions using an achiral standard catalyst $\langle RuCl_2[P(C_6H_5)_3]_3\rangle$ were remarkably enhanced when the reaction was carried out in the presence of an alkaline base and ethylenediamine in 2-propanol. In an enantioselective version of this approach, a chiral BINAP ligand was combined with a chiral diamine ligand in conjunction with ruthenium(II) and in the presence of KOH [36]. Generally, high TON and TOF values can be achieved in the asymmetric hydrogenation of ketones (indeed, TON values of AH catalysts are the highest reported in the context of asymmetric catalytic reductions of ketones). However, the rate is highly sensitive to the pressure of hydrogen. For example, hydrogenation at an S/C of 500 at 1 bar gave a TOF of 880 h^{-1}, whereas a TOF of 23 000 h^{-1} was obtained at an S/C of 10 000 at about 50 bar [34]. In a more recent study, base-free conditions were successfully applied when a modified BINAP-based catalyst, trans-RuH(η^1-BH$_4$)(binap)(1,2-diamine), was used [37].

2.3.2 Asymmetric Hydroboration

The application of boranes to the enantioselective reduction of carbonyls was first described in the late 1960s [38], but yielded only poor results until in 1979 Johnson used β-hydroxysulfoximine with diborane gas [39],

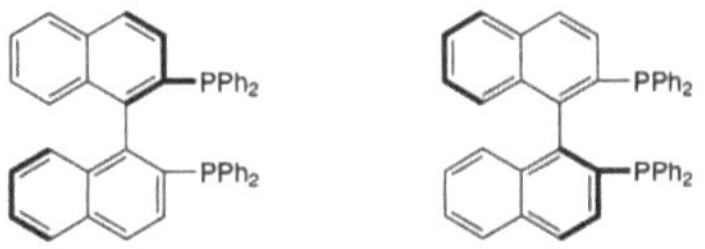

2.3 Enantioselective Catalytic Hydrogenation of Ketones

(a) R-BINAP   (b) S-BINAP

Figure 2.1: The BINAP ligand.

and in 1981 Hirao and co-workers used chiral amino alcohols with $BH_3 \cdot THF$ [40] to reduce prochiral ketones to chiral alcohols with optical yields of up to 82 % and 60 %, respectively. These approaches, however, were not catalytic, but the latter especially was seminal for later developments in the field. Through further studies by the group of Itsuno and the mechanistic investigations performed by Corey, Bakshi, and Shibata, the catalytic properties of oxazaborolidines were discovered [41, 42], and a novel analog known as the CBS-catalyst (after the inventors' names) was developed [43]. Scheme 2.1 shows the original CBS system with $BH_3 \cdot THF$ as the reducing agent, providing nearly quantitative yield in the transformation of acetophenone to (R)-1-phenylethanol (via boron enolates, which require acidic workup) with up to 97 % ee. Further investigations led to a wide range of chiral oxazaborolidine-based catalysts and reducing agents applicable to the highly enantioselective reduction of different ketones even on an industrial scale [41]. More recently, borohydrides such as $NaBH_4$ and $LiBH_4$ were successfully applied to enantioselective hydroboration, and effective metal catalysts have been developed [44].

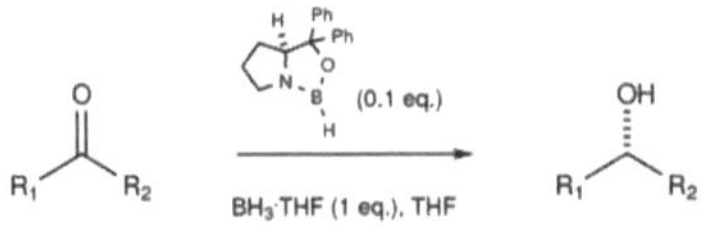

Scheme 2.1: The CBS reduction reaction.

2.3.3 Asymmetric Hydrosilylation

There are several ways to use hydrosilanes for the hydrogenation of unsaturated compounds; enantioselective hydrosilylation reactions of carbonyl and imino groups, however, are catalyzed by chiral metal catalysts with only few exceptions [45]. Rhodium especially, but also titanium, copper, zinc and other metals, are applied in combination with chiral ligands containing most often nitrogen or phosphorus as donor atoms. Different mechanisms are possible for the H-transfer, depending on the metal center [46]. In either case, a silyl ether is formed when carbonyls are applied as substrates and has to be subjected to acidic hydrolysis to yield the desired chiral alcohol. The silane chosen as the source of hydrogen has a strong impact on the enantiomeric excess of the products in the process. Although highly active and selective rhodium and copper-catalyzed asymmetric hydrosilylations have been developed, widespread industrial application has not yet been established [46].

2.3.4 Asymmetric Biocatalytic Reduction

Biocatalytic approaches are distinguished by their often extremely high (enantio)selectivities, mild reaction conditions, and harmless reagents. Nonetheless, the general instability of biocatalysts — which can be isolated enzymes or whole microorganisms — to thermal and mechanical stresses and pH variations limits their use. Additionally, development cycles are long, modifications difficult, and the number of commercially available biocatalysts is rather small [47]. Unlike several chemo catalysts which have a broad range of application (privileged catalysts), biocatalysts are usually effective for only one type of substrate [48]. In enzyme-catalyzed reductions of carbonyl groups, the hydride, which is transferred to the carbonyl C-atom, is delivered by a coenzyme; the reduced form of nicotinamide adenine dinucleotide (NADH) or the corresponding phosphate (NADPH) are commonly applied. Various methods and hydrogen sources are in use for the in situ regeneration of the coenzymes, which is an important issue because NAD(P)H is too expensive to be stoichiometrically employed [49]. Early biocatalytic ketone reductions were performed with baker´s yeast (whole cell approach); more recent studies on the reduction of aromatic substrates such as acetophenone have been performed using both microbial and enzyme/coenzyme approaches [50, 51].

2.4 Asymmetric Transfer Hydrogenation

Asymmetric transfer hydrogenation (ATH) is the stereoselective "reduction of an organic substrate by transfer of dihydrogen eqivalents from a suitable donor" [52]. Additionally, the general use of the term ATH implies that a catalytic step is decisive for both hydrogen transfer and chiral induction. As it is an operationally simple, mild, and efficient method for the generation of chiral substances from prochiral substrates, it supplements the afore-mentioned methods and may provide certain advantages. ATH is generally applicable to carbonyls, imines and C-double bonds, but it is particularly and most successfully used for the synthesis of non-racemic secondary alcohols from prochiral ketones [53, 54]. This is also the focus of the following considerations on different aspects of this method.

2.4.1 Historical Background

Hydrogen transfer reactions have been known since 1903 when Knoevenagel and Bergdolt reported the formation of 1,4-dihydro-dimethyl cyclohexa-2,5-dienecarboxylate by heating a mixture of dimethyl 1,4-cyclohexanedicarboxylate and dimethyl terephthalate in the presence of palladium [55]. The application to carbonyls was independently reported by Meerwein and Schmidt [56], Ponndorf [57], and Verly [58] more than twenty years later (MPV reduction, after the inventors´ names). They demonstrated that carbonyls can be converted into alcohols in the presence of superstoichiometric amounts of aluminum alkoxides in solution of easily oxidizable alcohols such as 2-propanol (Scheme 2.2). The reverse reaction was described later by Oppenauer [59].

The first "asymmetric catalytic" transfer hydrogenation reaction was reported by von Doering and Young in 1950 [60]. They reduced ketones by using racemic aluminum alkoxides in an excess of chiral alcohols and provided evidence for the hypothesis that the hydrogen transfer proceeds via a six-membered transition

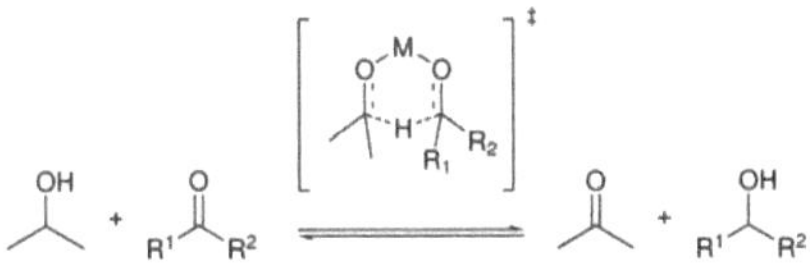

2.4 Asymmetric Transfer Hydrogenation

Scheme 2.2: The MPV reduction and Oppenauer oxidation.

state (cf. Scheme 2.2). Mitchell, Henbest, et al. reported the first example of a transition metal-catalyzed transfer hydrogenation using iridium complexes in 2-propanol [61, 62]. Some years later, Sasson and Blum applied a dichloro(tristriphenylphosphine)ruthenium(II) complex as catalyst for the transfer hydrogenation of unsaturated carbonyls [63], and subsequently published an extensive and seminal kinetic study [64]. In this study, they reported an acceleration in the rate of the reduction of unsaturated compounds in the presence of a base as co-catalyst. Nevertheless for most such reactions, high temperatures were required until Bäckvall and Chowdhury reported the ruthenium-catalyzed transfer hydrogenation of ketones under significantly milder conditions [65]. Just as in the field of asymmetric hydrogenation (see Section 2.3.1), the development of ATH since the early 1990s is linked to the name Noyori. His group introduced complexes based on (mono-sulfonated) diamine and β-aminoalcohol ligands, which still are among the most successful catalysts in ATH, and provided a number of seminal contributions regarding the mechanistic aspects of this type of reaction (see below).

2.4.2 Hydrogen Donor Systems

Arguably the most often employed hydrogen donor in ATH reactions is isopropyl alcohol (IPA = 2-propanol), frequently used with an inorganic base or an alkali metal alkoxide as co-catalyst to promote the increase of the concentration of 2-propoxide and/or to facilitate the generation of the true catalyst [66, 67]. Other alcohols have also been used, but were less efficient since — at least for catalysts based on mono-tosylated diamine or amino alcohol ligands — the order of reactivity has been determined to be 2-propanol > ethanol > methanol [68, 69]. This might be due to facile decarbonylation of primary alcohols (especially methanol), which causes catalyst poisoning [65]. Further hydrogen donors include formic acid, mostly applied as an azeotropic 5:2 mixture with triethylamine (TEAF), or formates used in aqueous solutions. The main characteristics of each hydrogen donor system are summarized in Table 2.2.

IPA is a suitable solvent for most substrates and catalysts, and it is easily disposed, economical, and (relatively) innocuous [66]. A major drawback of its use in ATH reactions is the reversibility of the conversion to acetone, affecting both the conversion of the substrate and the enantiomeric excess of the product. During the course of transformation of the substrate, the rate of the reverse reaction increases, and the ratio of the enantiomers comes under thermodynamic control, effecting erosion of the enantiomeric purity [54]. In order to disable the reverse reaction, IPA is mostly used in excess (as a source of hydrogen and solvent with low substrate concentrations); distilling off acetone is a further option. The back reaction is more easily avoided, however, when formic acid or formates are used as the hydrogen donors, because carbon dioxide is generated upon hydrogen transfer and can simply be vented. Nevertheless, the use of formic acid and

2 Fundamentals and State of the Art

aqueous solutions of formates is limited due to the instability of some catalysts in acidic media and water.

However, reactions carried out in water have attracted growing attention in recent years. Although there is some controversy on this subject [70], using water as the reaction medium is generally considered as a strategy toward "greener" approaches in chemistry since water is safe, sustainable, and economical [71, 72]. The insolubility of many organic compounds in water has often been assumed to be a hindrance for its use; in recent studies, however, a growing number of rate accelerations and improved selectivities have been reported for reactions performed in water [73], e.g., the asymmetric transfer hydrogenation of aryl ketones with significantly higher rates (and slightly lower ees) in HCO_2Na/H_2O than in TEAF [74]. The first ATH approaches in aqueous media were reported in 2001. Williams et al. added water to reactions carried out in IPA with water soluble ruthenium [75] as well as rhodium and iridium [76] catalysts, with the aim of developing biphasic systems and supported liquid phase catalysts. Chung et al. used the pure sodium formate/water system as a medium for the enantioselective reduction of aryl ketones. Additionally, they performed a reuse of the catalyst, which remained in the aqueous phase when the product was extracted with hexane [77]. An overview of the research in the field of ATH operations carried out in water has been given very recently by Wu and Xiao [78, 79].

Table 2.2: Commonly used hydrogen donor systems.

	2-propanol	Formic acid	Sodium formate/ water
Solvent	2-propanol	organic solvents	water
Promotor	inorganic base	triethylamine	none
Drawbacks	reversibility of the reaction	too harsh for some catalysts	solubility of reactants
Indication	IPA	HCO_2H; TEAF when used with triethylamine	HCO_2Na/H_2O

2.4.3 Catalysts

2.4.3.1 Metals

Transfer hydrogenation reactions are — with very rare exceptions — catalyzed either by main group metal alkoxides or by transition metal complexes. The great advances in asymmetric approaches in the last 30 years, however, have been made using the latter. The most often employed transition metal is ruthenium, followed by iridium and rhodium [66]. Despite their enormous success in an ever growing number of applications, these three metals have the drawback of high cost and toxicity. Thus, some research groups are currently investigating alternatives, and interesting results have been reported.

Furthering the work of Kagan et al. [80], Evans and co-workers reported an asymmetric version of the MPV reduction using a chiral tridentate samarium complex as catalyst [81]; aryl methyl ketones were able to be reduced with high enantioselectivity (up to 97 % ee) and high isolated yields (up to 96 %). List and Yang used a copper-bisoxazoline complex as chiral catalyst and Hantzsch esters as synthetic NADH-analogous hydrogen donors for the enantioselective transfer hydrogenation of α-ketoesters in chloroform [82]. This

2.4 Asymmetric Transfer Hydrogenation

approach, however, provided *ee* values above 90 % only at a low temperature (-25 °C) with prolonged reaction periods. Iron complexes have been successfully used by the groups of Beller [83, 84] and Morris [85] (cf. Section 4.1.1).

2.4.3.2 Ligands

A great number of chiral chelating ligands (chelate from ἡ χηλή, claw or pincer), several successful examples of which are depicted in Figure 2.2, have been reported in the context of ATH. As donor atoms, nitrogen, oxygen, phosphorus, and rarely sulfur or others are employed. The multidenticity of these ligands — belonging to one of the categories *bidentate*, *tridentate*, or *tetradentate* — is an important factor which influences the complex stability. Gladiali further distinguishes between "anionic" and "neutral" ligands, since possessing or not a protonated donor centre of appropriate acidity, e. g., tosyl-NH in monosulfonated diamines and OH in amino alcohols (see below) [54], "is crucial for enabling an outer or an inner sphere mechanism in H-transfer" [66]. However, amino alcohols and mono-sulfonated diamine ligands possess two protic donor atoms, amine-N/alcohol-O and amine-N/amide-N, respectively. The alcohol and amide groups of these ligands deprotonate upon metal insertion (to build the precatalyst) and act as anionic donor units, while deprotonation of the amino groups occurs in a second step under reaction conditions (to build the active catalyst and during the course of H-transfer, cf. Section 2.4.4). An important feature of these "anionic" ligands is thus the "NH effect", the interaction between amine-H and substrate-O in the ATH of ketones [67]. This effect, however, has also been observed with "neutral" ligands containing secondary amines, e. g., the tetradentate diphosphine/diamine (PNNP) ligand (Figure 2.2 c), which yielded a significantly more active catalyst than the diimine analog [86, 87]. Laue therefore suggested that the active catalyst contains one deprotonated secondary amine [88], and thus follows a pathway which is analogous to that of catalysts made from "anionic" ligands [89].

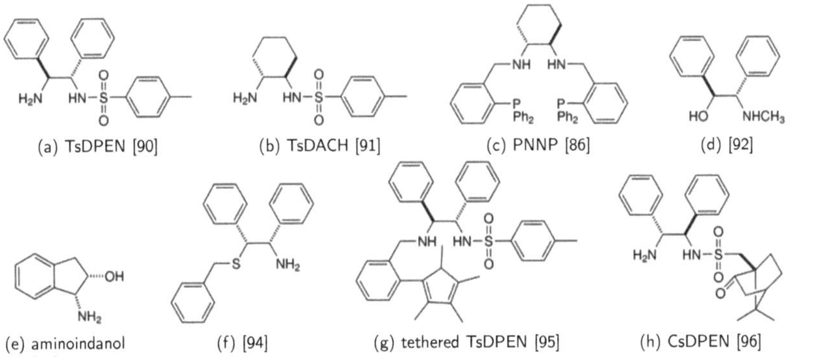

(a) TsDPEN [90]   (b) TsDACH [91]   (c) PNNP [86]   (d) [92]

(e) aminoindanol [93]   (f) [94]   (g) tethered TsDPEN [95]   (h) CsDPEN [96]

Figure 2.2: Ligands successfully used in ATH operations.

Complexes made of the *p*-toluenesulfononyl-diphenylethylenediamine (TsDPEN) ligand (Figure 2.2 a) and ruthenium-η^6-arene precursors (Ru-η^6-arene-TsDPEN, henceforth referred to as Ru-TsDPEN) are considered the most versatile catalysts in ATH, providing high enanatioselectivity with a variety of substrates [54]. Variants of the TsDPEN ligand, however, have given superior results in certain applications, e.g., the use of diaminocyclohexane (DACH) instead of diphenylethylenediamine (Figure 2.2 b), the tethering of a tetramethyl cyclopentadienyl moiety, which acts as an ancillary η^5-ligand (tethered TsDPEN, Figure 2.2 g), or the substitution of the tosyl for a camphorsulfonyl (Cs) moietey (Figure 2.2 h). Some β-aminoalcohol ligands such as 2-methylamino-1,2-diphenylethanol (Figure 2.2 d) in conjunction with [RuCl$_2$-η^6-C$_6$Me$_6$]$_2$ showed a higher activity than Ru-TsDPEN, but lagged behind in terms of enantioselectivity [92]. Experimental and theoretical studies of aminoalcohol-based catalysts have revealed the great impact of the ancillary arene ligands on the catalytic performance [67, 92, 97, 98]. Thus, both the activity and the enantioselectivity of the catalyst can be enhanced by substituting H-atoms of the η^5-cyclopentadienyl units — used in conjunction with rhodium(III) and iridium(III) — and η^6-arenes — used in conjunction with ruthenium(II) — for alkyl groups.

2.4.4 Mechanistic Aspects

For metal-catalyzed hydrogen transfer reactions involving carbonyl molecules, two basic mechanistic routes have been proposed: the *direct hydrogen transfer* and the *hydridic route* [99]. The first has been presumed to occur predominantly with main group metal catalysts; the latter is presumed for transition metal catalysts and can be further divided into monohydridic and dihydridic routes [100].

2.4.4.1 Direct Hydrogen Transfer

In catalytic reactions following this pathway, a hydride is transferred directly from the metal-coordinated reductant to the simultaneously coordinated oxidant via a concerted six-membered ring transition state, as depicted in Scheme 2.2 (see above). The mechanism has been supported by experimental as well as computational results for MPV reduction and Oppenauer oxidation in the classical version applying aluminum alkoxide as a promotor, or using aluminum and other non-transition metals as catalysts [101–104]. Some examples of transition metal catalyzed reactions, however, are proposed to follow an analogous pathway [94, 105].

2.4.4.2 Hydridic Route

In contrast to the direct pathway, hydridic routes imply a metal-hydride transition state where one or two hydrogen atoms are coordinated to the metal before being transferred to the substrate. Accordingly, a division into *monohydridic* and *dihydridic* reaction routes has to be made [100]. Rhodium and iridium catalysts generally favor the mono-hydride pathway, whereas the mechanism of the hydrogen transfer mediated by ruthenium catalysts depends on the ligands [106]. Additionally, reactions that follow the monohydridic path can proceed via an *inner sphere mechanism* or an *outer sphere mechanism*, depending on whether the ligand is neutral or anionic, respectively [66], or depending on whether there is an "NH effect" or not (cf. Section 2.4.3.2).

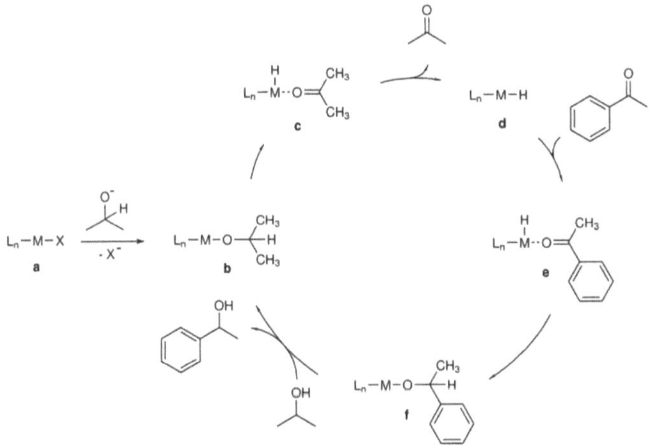

2.4 Asymmetric Transfer Hydrogenation

Scheme 2.3: Inner sphere mechanism.

Monohydridic Route A characteristic of metal mono-hydrides is that they are only formed by carbon-bound hydrogen atoms. Remarkably, these hydrogen atoms keep their identity, i.e., when alcohols are used as hydrogen donors, the alcohol´s C-bound H-atom forms the metal hydride and is then transferred to the carbonyl-C, whereas the alcohol´s O-bound H-atom is independently transferred to the substrate´s carbonyl-O [107]. In reactions that proceed in the **inner sphere** of the catalyst, a metal alkoxide is formed by the hydrogen donor and the catalyst [53]. Scheme 2.3 depicts as an example the reduction of acetophenone using 2-propanol as the hydrogen source and a transition metal complex as precatalyst (Sch. 2.3 **a**; X = anionic ligand, typically a halide; L = supporting ligand) [69]. After formation of the metal alkoxide (Sch. 2.3 **b**) via displacement of X by 2-propoxide in a basic medium, successive β-elimination and elimination of the acetone yield the metal hydride (Sch. 2.3 **b** → **d**). Linkage of acetophenone, followed by migratory insertion, then yields the new metal alkoxide (Sch. 2.3 **d** → **f**). The catalytic cycle is completed by ligand exchange and proton transfer yielding the metal propoxide (Sch. 2.3 **b**) and 1-phenylethanol as the product.

In contrast, there is no metal alkoxide formation in reactions that take place in the **outer sphere** of the catalyst. The hydrogen transfer instead occurs via a transition state where the substrate interacts with the catalyst via hydrogen bonds and dipolar interactions, without coordination to the metal center. A further subdivision of the outer sphere mechanism has been made by the proposal of a *stepwise* version in addition to the *concerted* mechanism, and by the proposal of a bimetallic pathway.

The concerted outer sphere mechanism — also known as metal-ligand bifunctional mechanism — is generally accepted for the ATH of ketones in the presence of Ru-η^6-arene complexes of mono-sulfonated diamines or β-amino alcohols, combined with an alcoholic hydrogen source [67, 69, 98, 108, 109]. Scheme 2.4 presents an example for this mechanism, which was first proposed by Noyori and co-workers in 1997 [68]. For both the formation of the precatalyst (Sch. 2.4 **a**) in situ and the formation of the active 16 e complex

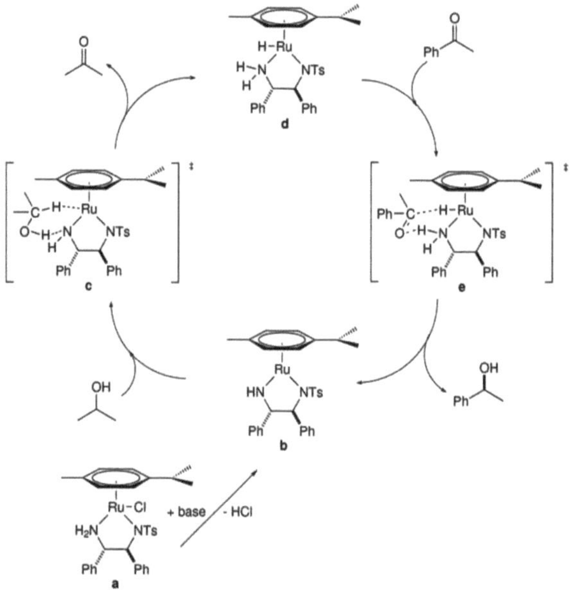

2 Fundamentals and State of the Art

Scheme 2.4: Concerted outer sphere mechanism.

(Sch. 2.4 **b**) via HCl-elimination, a base is required in 2-propanol solution; it is, however, not required as a co-catalyst, since the amine-N of the ligand acts as a basic center [110]. The transfer hydrogenation of the substrate catalyzed by the 18 e species (Sch. 2.4 **d**) involves a concerted transfer of proton and hydride (Sch. 2.4 **e**) received from the hydrogen donor (see Sch. 2.4 **c**). The precatalyst (Sch. 2.4 **a**) as well as the active catalyst (Sch. 2.4 **b**) and the reactive intermediate (Sch. 2.4 **d**) were isolated, and the molecular structures were confirmed [68]. A non-concerted but rather stepwise outer sphere mechanism, however, has been proposed, e. g., by Adolfsson et al. for the use of a ruthenium pseudo-dipeptide complex in the ATH of aryl ketone substrates performed in 2-propanol in the presence of lithium chloride [111].

Findings for the ATH of acetophenone in an aqueous/organic solution of excess sodium formate catalyzed by Ru-TsDPEN indicated a catalytic cycle similar to the concerted mechanism in 2-propanol (Scheme 2.5, cf. Scheme 2.4), as reported by Xiao, Liu, and co-workers [112]. When water is present, it acts as the base in the in situ formation of the precatalyst (Sch. 2.5 **a**). The addition of sodium formate provides the hydride species (Sch. 2.5 **c**) presumably via formation of a formato complex (**b**, species that could not be isolated). The hydrogen transfer to the substrate probably proceeds via a slightly modified transition state (Sch. 2.5 **d**); water significantly accelerates the reaction and — as supported by DFT calculations — is believed to alter the mode of hydrogen transfer from concerted to stepwise. As water acts as the source of protons, the reaction medium becomes increasingly basic with increasing conversion. A rate law consistent

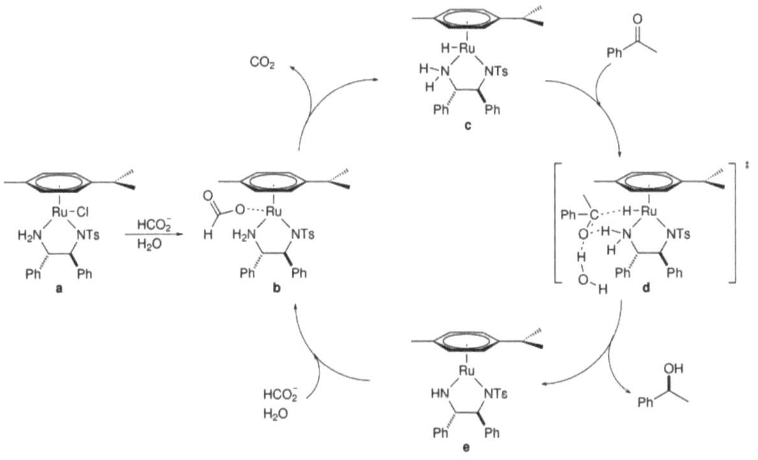

2.4 Asymmetric Transfer Hydrogenation

Scheme 2.5: Proposed outer sphere mechanism in aqueous media.

with the mechanistic consideratons and experimental kinetic results—a linear relationship relative to both the catalyst and the substrate concentration in a certain range—has been developed under the assumption that the coordination of formate and the decarboxylation are reversible and equilibrated prior to hydrogen transfer (Equation 2.4.1).

$$r = \frac{K_1 K_2 k[cat][HCO_2^-][acp]}{K_1 K_2 [HCO_2^-] + [CO_2][OH^-]} \quad (2.4.1)$$

Furthermore, the term $[CO_2][OH^-]$ was expected to be small since emerging carbon dioxide would react with the hydroxide ions released from the water molecules to form bicarbonate. Thus, when

$$[CO_2][OH^-] \ll K_1 K_2 [HCO_2^-]$$

is true, Equation 2.4.1 is reduced to Equation 2.4.2.

$$r = k[cat][acp] \quad (2.4.2)$$

On the basis of this rate law, the rate constant k was calculated at different temperatures in order to determine the *enthalpy of activation* ($\Delta H^\ddagger$) and *entropy of activation* ($\Delta S^\ddagger$) from an Eyring plot. As can be seen from Equation 2.4.4, which is the linear form of Equation 2.4.3, the slope of a plot of $\ln \frac{k}{T}$ against $\frac{1}{T}$ allows for the calculation of $\Delta H^\ddagger$, whereas $\Delta S^\ddagger$ can be calculated from the intercept [113]. The values of the activation parameters, especially $\Delta S^\ddagger$, are routinely used for mechanistic interpretation. Xiao, Liu, and co-workers reported a $\Delta H^\ddagger$ of 12.8 kcal mol^{-1} (53.6 kJ mol^{-1}) and a $\Delta S^\ddagger$ of -25 cal K^{-1} mol^{-1} (-105 J K^{-1} mol^{-1}) for the conversion of acetophenone [112]. From the large negative value of $\Delta S^\ddagger$, a well-ordered transition state

2 Fundamentals and State of the Art

(Sch. 2.5 **d**) was deduced.

$$k = \frac{k_B T}{h} \cdot e^{-\frac{\Delta H^{\ddagger}}{RT}} \cdot e^{\frac{\Delta S^{\ddagger}}{R}} \tag{2.4.3}$$

$$\ln\frac{k}{T} = -\frac{\Delta H^{\ddagger}}{R} \cdot \frac{1}{T} + \ln\frac{k_B}{h} + \frac{\Delta S^{\ddagger}}{R} \tag{2.4.4}$$

Dihydridic Route It has been observed in experiments with deuterated alcohols that hydrogen atoms transferred in ruthenium-dichloride-catalyzed reactions in the presence of a base lose their identities [107, 110]. Thus, a further mechanism has been proposed in hydrogen transfer reactions: the dihydridic route. In this mechanism, both the alcohol's O-bound and the α-C-bound H-atoms are transferred to the metal and then transferred from the metal to the ketone.

2.4.5 Applications and Industrial Impact

Asymmetric transfer hydrogenation has emerged from a method of mere academic interest to one that is used for the commercial production of chiral compounds [114]. While data about implemented commercial processes are difficult to obtain, and thus little is known about the number and details of such processes, the increasing interest in technical applications of this type of chemical operation is evidenced by the growing number of scale-up studies and tests of industrially relevant substrates by both academic and industrial researchers [20, 115, 116]. These studies demonstrate that a wide range of highly functionalized substrates can be employed in ATH operations even on a technical level. At the same time, it is apparent that each process is unique regarding the applied catalyst, reaction conditions, solvent, hydrogen donor system, etc., and great effort is required to find the optimal solution to a given problem. In particular, the transfer of the know-how developed under simplified laboratory conditions to technical applications remains challenging [19]. The predominant functional group reduced in ATH reactions on the pilot scale is the carbonyl group; imines, however, have also been employed. Figure 2.3 shows examples of chiral compounds for the production of pharmaceuticals which have been synthesized making use of ATH.

- In 2003, Merck researchers reported on the development of an ATH process for the manufacture of (R)-3,5-bistrifluoromethylphenyl ethanol (Figure 2.3 a), an intermediate in the synthesis of aprepitant (a pharmaceutical sold under the brand name *Emend*) [117]. The reaction was performed using a cis-aminoindanol-Ru(p-cymene) complex (cf. Figure 2.2 e, Section 2.4.3.2) as catalyst in the IPA/base system on a multi-10 kg scale. The chiral alcohol was obtained with an *ee* of 91 %, which had to be improved by further purification.

- A process including an ATH step, designed to avoid the resolution of a racemate and attendant waste, were developed for production of the drug *Diltiazem* [116]. The key building block cis-(2S)-lactam (Figure 2.3 b) was produced on a multi-liter scale by using an Ir-CsDPEN catalyst (cf. Figure 2.2 h, Section 2.4.3.2) at an S/C of 2 000 in a biphasic medium of water/isobutyl acetate with TEAF as the

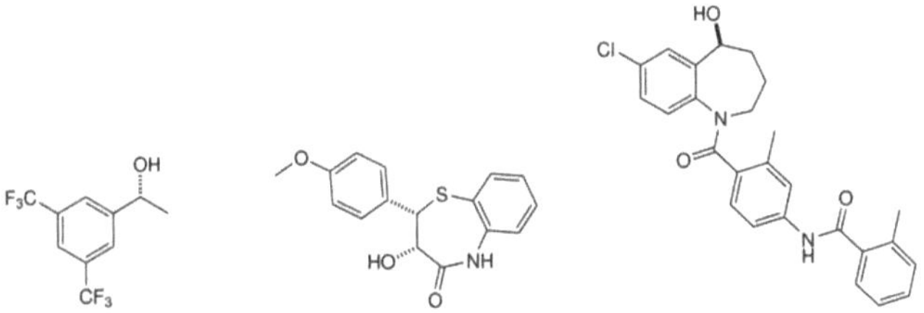

2.4 Asymmetric Transfer Hydrogenation

(a) (*R*)-3,5-bistrifluoro-methylphenyl ethanol

(b) *cis*-(2*S*)-lactam

(c) (*S*)-OPC-41061

Figure 2.3: Intermediates produced via ATH in scale-up studies.

hydrogen source. The product was obtained with an overall *ee* of more than 99 % while generating half the waste of the traditional route.

- Researchers at Otsuka studied the production of the Vasopressin V2 Receptor Antagonist OPC-41061 (Figure 2.3 c) on the laboratory scale (∼6 g) [118]. They used a Ru-TsDPEN catalyst (cf. Figure 2.2 a, Section 2.4.3.2) in IPA/KOH, and obtained the product in 99 % yield and 89 % *ee*, which was further improved via crystallization. This appraoch proved more efficient than asymmetric reduction methods using borane reagents.

A further indicator of the increasing commercial interest in ATH reactions is the growing number of commercially available catalysts. Under the name *CATHy*^{TM}, Avecia has developed catalyst kits, which are available from Strem Chemicals Inc. [119] and which contain aminoindanol and TsDPEN ligands as well as rhodium and iridium precursors to prepare the required ATH catalyst in situ. The TsDPEN ligand as well as a derivative (FsDPEN) are available from Takasago [120], and a tethered version of the η^6-arene-Ru-TsDPEN complex is available from Johnson Matthey [121].

2.4.6 Immobilization

As mentioned above, a major drawback of using homogeneously dissolved molecular catalysts in chemical production processes is their separation from reactants and products after the reaction has finished. To overcome this problem, a multitude of methods for the immobilization of catalysts has been developed, mainly in the context of asymmetric reactions [122–127]. To comprise all approaches, a definition of the term *immobilization* must be quite wide, for example "any strategy to facilitate catalyst separation and recycling". Catalyst immobilization strategies can be classified, e. g., by the state of the immobilized system under reaction conditions (solid or dissolved), by the support material applied (organic or inorganic), or by the method of immobilization (covalent binding, entrapment, micellar embedment, etc.). However, a complete discussion is outside the scope of this thesis, and two of the major categories — covalently bound catalysts and non-covalently bound catalysts — were built by considering the number of the respective applications rather than systematically. The most common techniques are discussed below, with recent examples in the field of asymmetric transfer hydrogenation referenced for each.

The starting point for the formation of an immobilized asymmetric catalyst is usually a chiral ligand or the corresponding metal complex, which has previously proven efficient in non-supported operations. The covalent binding of catalysts, either via copolymerization or via covalent attachment (anchoring) to a support, is the most often applied method for catalyst heterogenization [123], but is suitable for building soluble immobilized catalysts as well. Non-covalent immobilization includes approaches as different as the use of liquid supports, entrapment in porous solid supports, immobilization via adsorption, and others. Despite the great variety, approaches other than covalent attachment have gained less attention in ATH and studies on them are thus less in number.

Although significant advances have been made in the fields of both asymmetric catalysis and catalyst immobilization, there is apparently no use of immobilized chiral metal complexes on an industrial scale [3, 5, 123]. Blaser and Pugin have identified a number of requirements on heterogeneous asymmetric catalysts for commercial use, most of which have so far not been met [5]. As a result, most immobilized catalysts are not yet competitive compared to alternative methods such as enantiomer resolution and chiral pool synthesis, or homogeneous catalysis. The requirements on heterogeneous asymmetric catalysts include the variability (combinatorial approaches) and cost efficiency of their preparation (additional costs of the immobilized catalyst should be lower than those for the separation of the homogeneous catalyst or outweighed by other advantages such as catalyst reuse), their performance (selectivity, activity, productivity should be comparable or better than of the homogeneous catalyst, 95 % recovery and minimal metal leaching is required, cf. Section 2.2.2), and their handling (simple separation is required). Furthermore, the molecular weight of the heterogeneous catalyst should not exceed 10 kDa per mole of active site. Additionally, it is crucially important that it be available in technical quantities within the time frame of the process development [5]. In fact, availability is a problem even for testing on a laboratory scale; at least immobilized ATH catalysts are not yet commercially available [128] (see Section 5.4.2).

2.4.6.1 Immobilization via Covalent Binding

Independent of the procedure, covalent binding mostly requires the modification of the ligand structure to introduce at least one functional group (linker), and in some cases a certain distance between the linker and the catalytically active center has to be assured by a spacer. Furthermore, the applied support material or co-monomer, the length and flexibility of the spacer, the catalyst loading, etc. affect the accessibility of the active catalyst center and the final catalyst performance. In the following, covalent attachment and polymerization approaches are discussed separately. The first is distinguished from the latter by the existence of the support prior to the immobilization procedure; that is, a ligand or catalyst is attached to an existing support in order to form the immobilized catalyst. In case of (co)polymerization, though, the support is formed from the ligand (and the co-monomers) during the immobilization procedure.

Covalent Attachment Polywka et al. presented an early example for the heterogenization of ATH catalysts [129]. As depicted in Scheme 2.6, the ligand was prepared by coupling a modified tosyl moiety (containing a linker) with the chiral diamine backbone. Amino resins were used as solid supports to which the modified TsDPEN ligand was efficiently attached via peptide coupling. Metal insertion using

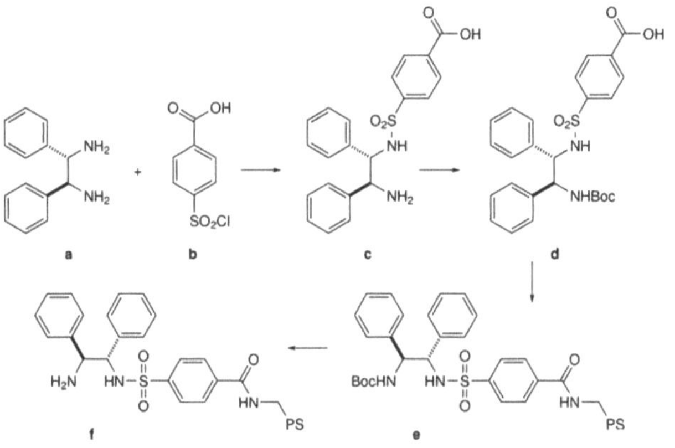

2.4 Asymmetric Transfer Hydrogenation

Scheme 2.6: Covalent attachment of a TsDPEN derivative to aminated polystyrene supports.

a [RuCl$_2$(p-cymene)]$_2$ precursor was performed in situ when the polymer-bound catalyst was used for the ATH of acetophenone in IPA and TEAF systems. Ee values of up to 99 % were achieved, but the reaction was slow, and the catalyst was nearly non-reusable.

A similar approach has recently been presented by Somanathan and co-workers [130]. Instead of a TsDPEN-based ligand, a modified TsDACH ligand (cf. Figure 2.2 b, Section 2.4.3.2) was coupled to aminomethylated polystyrene, and used in conjunction with [RhCl$_2$(Cp*)]$_2$ as a heterogeneous catalyst for the ATH of aryl ketones in water/sodium formate. An enantiomeric excess of 90 % and nearly complete conversion were observed, and the catalyst was recycled four times with only a slight drop in activity.

Van Leuwen, Reek, et al. reported the attachment of a trimethoxysilane-functionalized aminoalcohol ligand onto silica. The remaining silanol sites on the support were transformed into alkylsilane sites in a further step [131]. Both batchwise and continuous applications of the immobilized ligand and the [RuCl$_2$(p-cymene)]$_2$ precursor as catalyst were successfully performed in IPA. In a similar approach, the introduction of ethyltrimethoxysilane as a spacer/linker into the tosyl moiety allowed a TsDPEN derivative to be anchored onto different silica supports. The heterogenized ligand in conjunction with [RuCl$_2$(p-cymene)]$_2$ was successfully used for the ATH of aryl ketones in the organic TEAF solutions [132] as well as in aqueous media [133, 134].

Further heterogeneous silica-supported systems were reported in recent years [135–137]; one remarkable approach was provided by Li et al., who immobilized a Ru-TsDPEN catalyst in a siliceous mesocellular foam which was magnetized through the grafting of magnetic nanoparticles [138]. The use of cobalt nanoparticles as support for ATH catalysts was reported by Pericàs and co-workers [139].

An example for attaching a (pre)catalyst to a soluble support was given by Liese et al. [140]. Their "chemzyme approach" consisted of a PNNP-based ruthenium catalyst (cf. Figure 2.2 c, Section 2.4.3.2) which was linked to a polysiloxane chain (Section 4). The solubility of the polymer in 2-propanol was increased by additionally attaching polar tris(2-methoxyethoxy)vinylsilane, and the catalyst was successfully

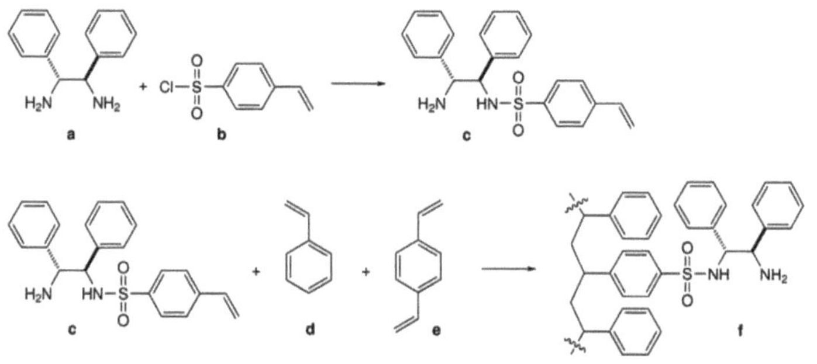

2 Fundamentals and State of the Art

Scheme 2.7: Crosspolymerization of a TsDPEN-based ligand monomer.

used in a continuously operated membrane reactor providing high reusability (see also Section 4.1.2).

(Co)polymerization Chiral polymers had been used for various purposes long before immobilized ATH catalysts gained interest [141]. Arguably the first example of an immobilization approach for ATH applications was reported by Lemaire et al. in 1994 [142]. Polyamides and polyureas were made from a chiral diphenylethylenediamine ligand in conjunction with terephthaloyl chloride and bis(1,4-isocyanatophenyl)methane, respectively. Although rather modest, the overall performance of the insoluble rhodium-polyurea catalyst was better than that of the non-immobilized rhodium-diamine in the ATH of acetophenone in 2-propanol/KOH.

In a variation on this approach, preformed rhodium complexes were employed for copolymerization, and **molecular imprinting** was applied in order to improve the enantioselectivity [143]. In fact, the templated copolymers yielded better results than the non-templated ones in terms of the enantioselectivity, but the overall performance was still not satisfactory [144]. Further examples of copolymerization combined with molecular imprinting were reported by Severin and Polborn, who applied a "styrene derivative" of the TsDPEN ligand [145–148].

The same modified TsDPEN ligand had been previously used by Lemaire for copolymerization with styrene and styrene/divinylbenzene to get both linear and crosslinked polymers (see Scheme 2.7), respectively, which were used with iridium and ruthenium precursors as catalysts for the ATH of acetophenone in IPA [149, 150]. This approach was seminal for further developments. The introduction of hydrophilic pendent groups into the polymer framework allowed the catalyst to be successfully applied to the ATH of ketones and imines in aqueous reaction media [151–153].

2.4.6.2 Non-Covalent Immobilization

The significant advantage of many immobilization methods not based on covalent binding is that a modification of the ligand is not necessarily required. For some of these methods, recent examples can be found in the context of ATH.

2.4 Asymmetric Transfer Hydrogenation

Liquid Supports As mentioned above, Chung et al. reported arguably the first liquid-supported ATH catalyst [77]. They used a modified proline amide ligand, which was made water soluble through the introduction of a fluoride group, and obtained promising results, although the activity dropped slightly upon recycling. In five consecutive runs with 94–95 % ee, the time until nearly complete conversion was achieved was prolonged from four to seven hours.

Fan, Gu, Chan, and co-workers reported the ATH of acetophenone mediated by unmodified Ru-TsDPEN in mixtures of polyethylene glycol (PEG) and different hydrogen donor systems. The best results in terms of recyclability and reactivity were obtained when a homogeneous mixture of PEG and sodium formate/water was used as the reaction medium. After each run, the product was extracted with hexane, while the catalyst remained in the PEG/water phase. The hydrogen donor was regenerated by the addition of formic acid, and up to 14 recycling experiments were performed with nearly stable ees (94–96 %) but with increasing reaction periods [154].

Micelles The first report on the use of surfactants in aqueous ATH was published by Chung et al. [155]. An unmodified proline-based ligand in conjunction with a ruthenium precursor was used as the catalyst for the reduction of various aryl ketones in sodium formate/water. Recycling of the catalyst was possible, but the performance after recovery was rather poor.

Zhu, Deng, and co-workers reported the use of Ir, Rh, and Ru-TsDPEN catalysts in the presence of micelle-forming surfactants in sodium formate/water [156]. Ru-TsDPEN, embedded in micelles formed from cationic surfactant cetyltrimethylammonium bromide, was successfully applied to six consecutive runs of the ATH of acetophenone, in which stable ees of 95 % were obtained but reaction periods increased (first run: 6 h to conversion >99 %; sixth run: 13 h).

Entrapment The encapsulation of a homogeneous Ru-TsDPEN catalyst in mesoporous silica via adjustment of the pore entrance size by silylation was demonstrated by Yang, Li, and co-workers [157]. The supported system was reusable multiple times in ATH reactions of aryl ketones in sodium formate/water, but the reuse suffered from increasing reaction periods, probably due to the loss of catalyst during the course of recovery. Using acetophenone as the substrate, the ee was stable at 92–93 % over six runs, the same enantioselectivity as achieved with the non-encapsulated catalyst. In a very recent study, this approach was significantly enhanced by tuning the microenvireonment of the catalyts within the nanocage using an amphiphilic silylation agent [158]. The modified cage showed a highly increased adsorption capacity for water, benzene, and negatively charged ions. The higher rates at stable ees achieved with this system were therefore ascribed to the ability of the cage to accumulate the reactants.

3 Support Materials Applied

Appropriate support materials for catalyst heterogenization purposes have to satisfy several conditions: chemical and mechanical stability, high specific surface area, high density of functionalities as well as accessibility and acceptable costs. Additional requirements depend on the specific application, for example a high or low hydrophilicity of the surface, or a special geometry.

To achieve the aim of this thesis, the development and study of highly reusable catalysts for ATH operations, immobilization via covalent attachment to solid supports was the method of choice. On the one hand covalent bonds provide the strongest type of linkage, thus impeding catalyst loss, and on the other hand solid supports have known material properties and most efficiently allow for facile quantitative separation. This required materials with the above-mentioned properties in addition to functional groups which allow for catalyst immobilization via standard coupling procedures. Such supports were supplied by PolyAn GmbH, a surface-technology-focused company located in Berlin. Special needs were met by the joint development of a new support type.

3.1 Material Choice and Preparation

PolyAn specializes in *Molecular Surface Engineering* (MSE) and provides custom-made, molecular-designed material surfaces and boundary layers, equipped with specific functions [159]. These materials are mainly used as supports for biomolecules for the application in medical diagnostics and for use in membrane-based separation processes. Other applications that have been studied include the use of surface-functionalized polymer chips (PCs) as supports for C–C-coupling catalysts [160]. These attempts were considered a starting point, and analogous PCs were chosen for the present study, i.e., aminated membrane slices made of polypropylene (Figure 3.1 a) and sinter chips made of polyethylene (Figure 3.1 b). Both types of polymer chips were readily available from PolyAn, but for several experiments which had to be conducted under

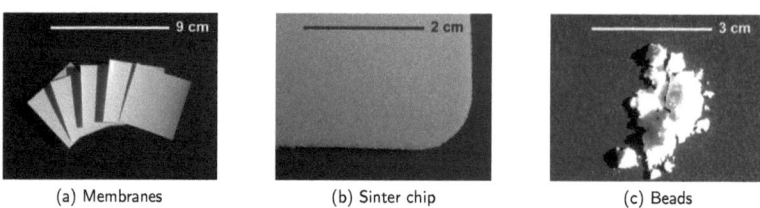

(a) Membranes (b) Sinter chip (c) Beads

Figure 3.1: Support materials applied.

highly reproducible conditions in a flask with mechanical stirring, another support version, namely beads, was required (cf. Section 5.2.1). Thus in a joint project, the MSE-based functionalization was adapted for ultra high molecular weight polyethylene (UHMW-PE) particles of 30 µm in diameter, generously provided by Ticona (Figure 3.1 c, micro particles henceforth denoted as polymer beads, PBs). While the details of the PolyAn technology cannot be published, some aspects of the material properties are discussed in the following.

3.1.1 Molecular Surface Engineering

Molecular Surface Engineering denotes the design of the chemical properties of material surfaces by utilizing copolymerization techniques (grafting from). The method is applicable to a wide range of base materials with different morphologies, whether inorganic like glass or organic like artificial polymers, and whether with planar or porous structures etc.

Hence, the applied base materials are modified by grafting a "matrix" with terminal functional groups. The advantage of MSE over other techniques is that the matrix is covalently bound to the base material, and that the chosen functional sites are mono-fractional. Both the functional groups as well as further properties of the matrix, e. g., the hydrophilicity–hydrophobicity balance, can be chosen from various options or ranges [159]. Additionally, due to the three-dimensional structure of the grafted matrices, comparatively high loadings of accessible functional groups are provided.

3.1.2 Material Properties

The materials applied in this study were composed of polypropylene (PP) or polyethylene (PE) base materials and the grafted polymer matrix, which contained both hydrophilic and hydrophobic compounds as well as terminal primary amines as functionalities (Figure 3.2). The amino groups were Boc-protected for chemical stability during storage. The balance of hydrophilicity and hydrophobicity of the polymer chips provided a sort of amphiphilic character, which had a positive effect on the reaction rate when the materials were used as catalyst supports for the reaction of organic substrates in polar media (see Section 5.2.2.1). Thus, further optimization of the matrix properties were not considered necessary, and beads were functionalized in an analogous manner using an adapted process.

The supports were manufactured with different loadings of amino groups, usually in the range of 10–30 $\mu mol\, g^{-1}$. The number of functional groups on the surface is related to the thickness of the grafted matrix, which can be controlled in the MSE process. A study of analogous polymer materials showed that the amount of detectable functional groups per surface area follows a saturation curve when plotted versus the layer thickness, since with increasing layer thickness there is a growing number of inaccessible groups. Reasonable values for the matrix thickness are in the range of 10 nm to 300 nm as determined via interferometry and other methods [161].

Accurately determining of the number of (accessible) amines is challenging, and different methods will provide different results. A reasonable method to quantify them is the coupling of (cleavable) compounds which absorb in the ultraviolet (UV) and visible (Vis) range, which is carried out as follows: after coupling of

3 Support Materials Applied

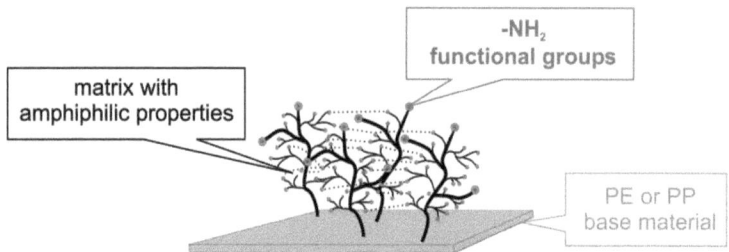

Figure 3.2: Schematic illustration of the support materials.

the specific compound, the supports are washed, the UV active part is cleaved, and finally the concentration, which is equal to the effectively accessed amino groups, is determined via UV absorption measurement. Another method is to use an accurately determined amount of the coupling reagent and measure the fraction which has not coupled to the surface, again via UV-Vis or fluorescence analysis. For this, different compounds (Fmoc-β-Ala-OPfp [Fmoc], ortho-phthaldialdehyde [OPA], Ponceau S) were applied, and a systematic deviation was found, which might primarily be due to the molecular size of the respective compounds. The same was observed when the amount of metal determined via inductively coupled plasma mass spectrometry or optical emission spectroscopy (ICP-MS/ICP-OES) of digested PB and PC supports was compared to the number of amino groups previously determined using Fmoc as the reference compound.

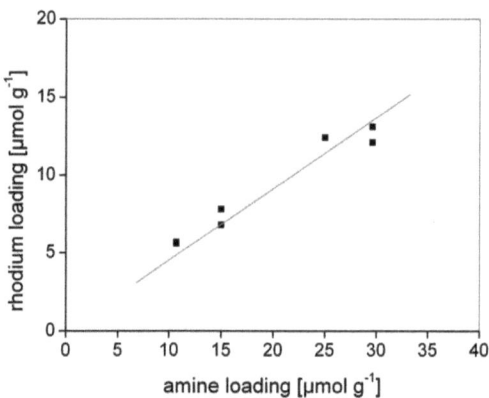

Figure 3.3: Loading with amino groups vs. loading with rhodium.

Figure 3.3 shows that there is a linear relationship in the investigated range, but that there is less coupling of the catalyst complex (see Section 5.1.2) than Fmoc coupling. The Fmoc/UV-Vis test was used as the standard procedure to determine of the amine loading, taking into account that absolute values might be somewhat deceptive. ICP-based analysis, on the other hand, was used to determine the catalyst loading, assuming that the amount of metal found is equal to the amount of catalyst.

Consequently, it became evident that a certain number of free amino groups were still present after catalyst immobilization. Tests of functionalized materials (carboxyl function) have revealed that the number of transformed groups is much smaller than that of non-transformed [161]. The free amines might have an additional effect on the material properties, but this (potential) effect could not be investigated in the present study. However, the mechanical stability of the polymer beads, which were required to ensure homogeneous suspension through vigorous mechanical stirring, was proven via optical determination of the size distribution. No significant deviation was observed in the particle size of unused beads and beads employed in an ATH experiment performed at 1 200 rpm and 60 °C.

4 Modification of a Ruthenium Catalyst

ATH ligands or (pre)catalysts which meet the requirements for immobilization are not commercially available. In academia, however, a number of pathways have been developed for the modification of various ATH catalysts with the aim of immobilization. Thus, the manufacturing of ligands or catalysts appropriate for immobilization could be managed either by following one of the routes described in the literature or by developing a new route. As described above, for the purpose of covalent attachment the ligand has to possess a suitable functional group and should provide high complex stability through multidenticity. A tetradentate diphosphine/diamine ligand, which has widely been used in the context of ATH, was chosen as a candidate for immobilization. Since none of the prior attempts at modification of this ligand were considered appropriate for the covalent attachment to aminated supports, a new strategy for the introduction of a linker was required.

4.1 Background

4.1.1 Homogeneous Applications

Diphosphine/diamine ligands had been studied for some time [162–164] before the N,N'-bis[2-(diphenylphosphino)-benzyl]cyclohexane-1,2-diamine (PNNP) ligand (Figure 4.1 a) and its diimine derivative were used in conjunction with ruthenium(II) (Figure 4.1 b) as (pre)catalysts for ATH reactions by Gao, Ikariya, and Noyori in 1996 [86]. The Ru-PNNP catalyst provided 97 % ee and 93 % yield (TON = 186, TOF = 27 h^{-1}) in the asymmetric transfer hydrogenation of acetophenone in IPA/base. Both the rate and enantioselectivity provided by the diimine-based version were significantly lower. Promising results, however, were obtained with cationic PNNP-based rhodium(I) complexes with different anions in the IPA/base medium [165]. Nevertheless, in terms of the overall performance of the ATH reaction, the neutral Ru-PNNP complex was favorable over the ionic rhodium version [166].

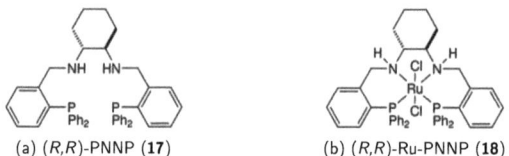

(a) (R,R)-PNNP (**17**) (b) (R,R)-Ru-PNNP (**18**)

Figure 4.1: Diphosphine/diamine ligand and ruthenium complex.

PNNP has also been used for ATH operations in conjunction with rhodium and iridium precursors in

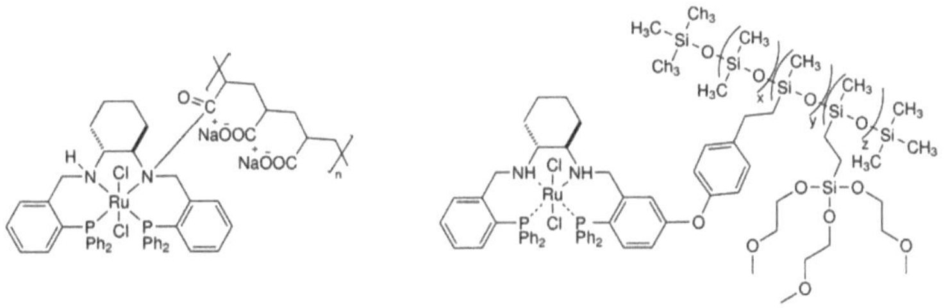

4.1 Background

(a) PAA-(*R,R*)-Ru-PNNP

(b) "PDMS"-(*R,R*)-Ru-PNNP

Figure 4.2: Immobilized ruthenium-PNNP complexes.

2-propanol without the addition of a base, and high yields and *ee* values were obtained with various aryl ketone substrates [167]. Morris et al. reported on the successful asymmetric transfer hydrogenation of ketones performed in the 2-propanol/base system using an iron complex containing the diimine version of PNNP as ligand [85]. More recently, this approach was applied to imines as substrates by Beller and co-workers [84]. The PNNP ligand has also been used in aqueous reaction media; Gao et al. reported the in situ generation of a PNNP-based iridium catalyst and its application to the ATH of various aryl ketones in sodium formate/water in the presence of phase transfer catalysts such as bis(triphenylphosphoranylidene) ammonium chloride (PPNCl) [168]. Further applications of complexes on the basis of diphosphine/diamine or diphosphine/diimine ligands for catalytic operations include asymmetric epoxidation and oxidation reactions [169–171].

Generally, the *R*-enantiomer of the product is formed in ATH reactions of prochiral aryl ketones when the *S,S*-version of the PNNP-based catalyst is applied, whereas the *S*-enantiomer is obtained when the *R,R*-catalyst is used. Acetophenone is usually employed as a benchmark substrate (see Scheme 4.1).

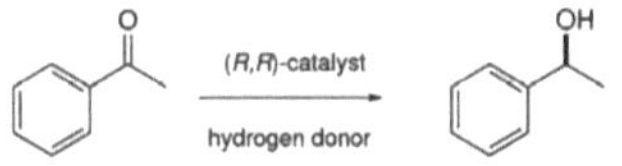

Scheme 4.1: ATH of acetophenone as the benchmark reaction.

4.1.2 Immobilization Approaches

Two pathways have been established to immobilize complexes based on **18** via covalent attachment; in both cases, soluble polymers were used as supports. Gao et al. presented an immobilized version which is characterized by simple preparation, since no modification is required to link one of the amines of the complex with poly acrylic acid (PAA, see Figure 4.2 a) [172]. The performance of the supported catalyst, however, was rather poor, displaying significantly prolonged reaction times compared to the free analog, decreased *ee* values, and poor reusability.

The other example was presented by Liese et al. [140]. Their "chemzyme approach" consisted of a modified version of **18** which was linked to a polysiloxane chain (Figure 4.2 b). The solubility of the polymer

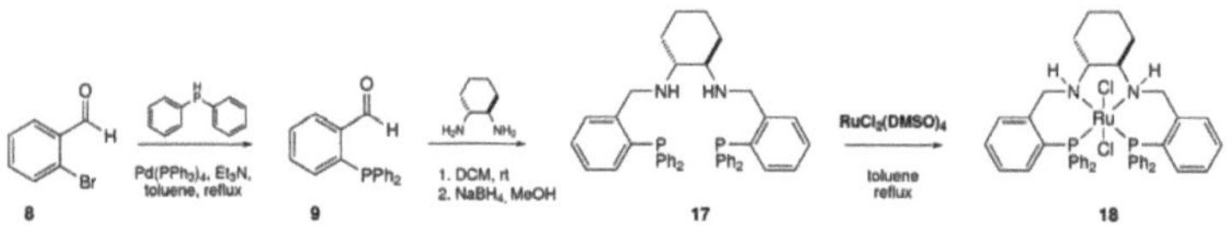

4 Modification of a Ruthenium Catalyst

Scheme 4.2: Preparation of the unmodified PNNP ligand and ruthenium complex.

in 2-propanol was increased by additionally attaching polar tris(2-methoxyethoxy)vinylsilane. A total of 12 steps were required for the preparation, and the resulting catalyst was successfully used in a continuously operated membrane reactor providing high reusability. A detailed description of the system performance was given, with reported STY values of up to 578 g L^{-1} d^{-1} (24 g L^{-1} h^{-1}), a TTN of 2 630, an *ee* of 91 %, and a TOF of 0.22 min^{-1} (13 h^{-1}) for the ATH of acetophenone in IPA/base. Furthermore, a detailed kinetic study of this system was reported by Greiner and co-workers [173].

4.2 Initial Studies

The high catalyst stability and reusability reported for the "chemzyme approach" as well as the high versatility of the PNNP ligand indicated by various studies — not only in the field of ATH — were encouraging signs that this system could be made applicable for the attachment to solid supports. Initial studies were undertaken in order to identify the strengths and weaknesses of different PNNP-based catalysts as well as potential strategies for synthetic modification.

4.2.1 Preparation and Use of Diphosphine/Diamine-based Catalysts

4.2.1.1 Ruthenium-Diphosphine/Diamine

The "original" PNNP ligand is prepared in a straightforward way by coupling two equivalents of 2-(diphenylphosphino)benzaldehyde (**9**) with either the *R,R*- or the *S,S*-version of diaminocyclohexane followed by subsequent reduction [165], as depicted in Scheme 4.2. Metal insertion using dichlorotetrakis(dimethyl sulphoxide)ruthenium(II) as precursor yields the Ru-PNNP complex in a further step [86]. In this study, only the *R,R*-version of the ligand (**17**) and the catalyst (**18**) was prepared. **18** was applied as homogeneous catalyst for the conversion of acetophenone to (*S*)-1-phenylethanol in the 2-propanol (hydrogen donor and solvent)/potassium propan-2-olate (co-catalyst) system. In addition to the studies of Gao and co-workers [86, 166] (cf. Section 4.1.1), a detailed description of this catalyst including thermodynamic and mechanistic considerations has been carried out by Laue [88]. He pointed out that a high enantiomeric excess is contrary to a high yield (cf. Section 2.4.2), resulting in a decreased *ee* of 92 % (intermediate maximum *ee*: 96 %) when the reaction is run to 98 % yield. Testing **18** confirmed a maximum excess of (*S*)-1-phenylethanol of 95–96 % and revealed the high sensitivity of the catalyst to moisture and air. Thus, a "blow off" of acetophenone to increase the final *ee* and the reaction rate was tested by applying an argon overflow in an open flask. Even though this approach was effective, it is, of course, not an optimal answer to the problem.

Table 4.1: Performance of Ir-PNNP.
Reaction conditions: acp 1.2 mmol, HCO$_2$Na 1.8 mmol, water 20 mL, (R,R)-PNNP 0.03 mmol, [IrHCl$_2$(COD)]$_2$ 0.013 mmol, toluene 2–3 mL, 50 °C.

Entry	Additive	Time [h]	Conversion [%]	ee[a] [%]
1	–	24	70	30
2	SDS (1 g)	8	<5	n. d.
3	Lutensol XA 60 (1 g)	24	80	54
4[b]	PPNCl (5 mol%)	47	99	62

[a] excess of the S-enantiomer, n. d. = not determined
[b] taken from [168]; acp 0.25 mmol, HCO$_2$Na 1.25 mmol, water 2 mL, (R,R)-PNNP 0.0055 mmol, [IrHCl$_2$(COD)]$_2$ 0.0025 mmol, 60 °C

4.2.1.2 Iridium-Diphosphine/Diamine

In order to test further applications, the PNNP ligand was used in conjunction with an iridium precursor as catalyst for ATH operations performed in aqueous media. Compound **17** and [IrHCl$_2$(COD)]$_2$ were dissolved in toluene or tetrahydrofurane (THF) and stirred for a moderate period of time before the hydrogen donor (sodium formate/water) and the substrate were added. The reaction could also have been performed via in situ generation of the catalyst in water, but an organic solvent was preferred since in water the catalyst tended to adhere to the stirrer tool. As reported by Gao and co-workers, the performance of the system without the addition of phase transfer catalysts was rather poor [168]. By employing excess surfactants, it was tested if an improvement in rate and enantioselectivity could be achieved. Furthermore, immobilization was attempted using the formation of micellar emulsions. As shown in Table 4.1, the anionic surfactant sodium dodecyl sulfate (SDS) almost completely inhibited the reaction whereas the non-ionic surfactant Lutensol XA 60 indeed enhanced rate and ee. Nevertheless, these initial attempts did not achieve the results reported by Gao et al. (Table 4.1, Entry 4).

Due to the rather low enantiomeric excess achieved with Ir-PNNP in sodium formate/water, the potential for application as an immobilized catalyst in ATH operations was limited. As pointed out in Section 2.2.1, ees in asymmetric catalytic processes should be >90 %, but might be acceptable if they exceed 80 %; less than 70 % ee, however, is too low. The focus of immobilization attempts therefore lies with the ruthenium catalyst. Nevertheless, an immobilized non-enantioselective catalyst might be interesting for other applications, and optimization of the reaction conditions might further improve the overall performance of the system.

4.2.2 Strategies for Modification

To achieve the aim of covalent attachment of the ligand or catalyst to the aminated supports (cf. Chapter 3), a modified, linker-containing version of the PNNP ligand had to be prepared. As a basis for the decision on how to implement the linker via a new synthetic route, the following considerations were made.

- The carboxyl group is the suitable linker which corresponds to the amino-functionalized supports since coupling can be performed via standard amide formation procedures.

4 Modification of a Ruthenium Catalyst

- The linker has to be introduced into the ligand as far as possible from the catalytic center.
- A certain distance between catalyst and support should be provided by a spacer.
- The compounds required for modification should be commercially available to keep the number of synthetic steps as few as possible.

Due to the striking results achieved with the polysiloxane-supported catalyst by Liese et al. (cf. Figure 4.2 b), the 5-position of compound **8** and **9** was considered the suitable site for the introduction of a linker/spacer moiety. The number of applicable, commercially available compounds, however, was limited, and 2-bromo-5-iodobenzoic acid (**3**) was chosen as the most promising candidate. Thus, apart from the transformation of the carboxyl to an aldehyde group as well as the introduction of diphenylphosphine in 2-position (cf. Scheme 4.2), the coupling of a C_4–C_6 chain carrying a carboxyl group would have to be performed as the key step.

Knochel, Cahiez, et al. reported the formation of organozinc halides which could be selectively coupled with aromatic iodides even in the presence of further functional groups [174]. Encouraged by these results, it was attempted to insert zinc into bromo-hexanoic acid *tert*-butyl ester and couple the intermediate zinc species with the model compound 1-bromo-4-iodobenzene. Unfortunately, the formation of the organozinc halide did not succeed, and another strategy had to be applied. The selective coupling of a further C_6-acid ester, namely 5-hexynoic acid *tert*-butyl ester (**2**), at the iodo functionality of 1-bromo-4-iodobenzene was effectively performed via Sonogashira coupling. Hence, the strategy for the preparation of the modified PNNP ligand entailed the coupling of an alkynyl acid with 2-bromo-5-iodobenzoic acid or a derivative, which after further steps would provide the linker-containing analog to compound **9**.

4.3 Preparation and Use of the Modified PNNP

4.3.1 Synthetic Pathway

The synthetic route was divided into three major sections, the first of which was considered to be the most challenging and accompanied by considerable losses. Almost all reactions required strictly anhydrous conditions and an inert atmosphere. Scheme 4.3 shows the synthetic steps required for the preparation of the first key intermediate, the "modified diphenylphosphinobenzaldehyde" (**7**). Since the designated linker — the carboxyl group of 5-hexynoic acid (**1**) — might have hindered further synthetic steps and work-up procedures, a protective group was required. The introduction of the easily cleavable "*tert*-butyl" group into **1** was achieved via Steglich esterification to yield **2**. Commercially available compound **3** was reduced to the corresponding alcohol **4** by treatment with borane dimethyl sulfide complex. Both intermediates **2** and **4** were linked via a palladium-catalyzed Sonogashira coupling reaction, yielding compound **5**. By treating **5** with Dess–Martin periodinane, aldehyde **6** was formed, which in the next step was coupled with diphenylphosphine to form key compound **7**. Due to losses on all stages, an overall yield of **7** of 16.5 % (not including **3** → **4**) was obtained.

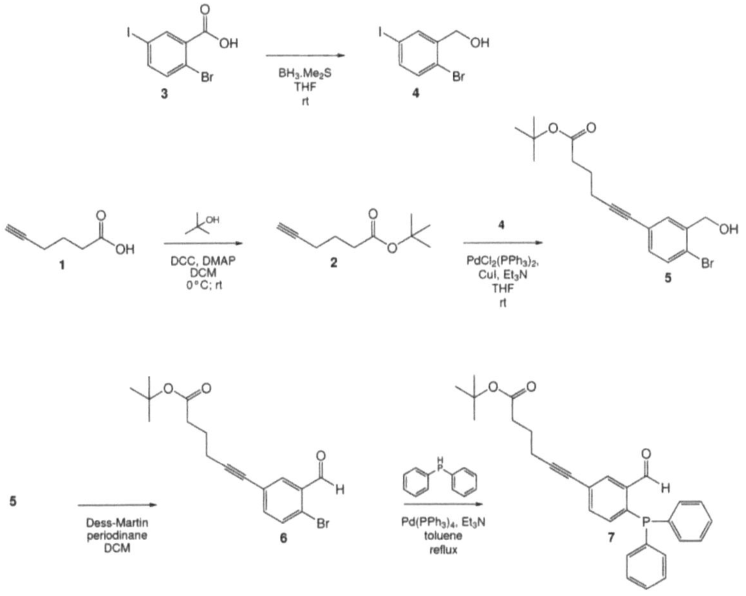

4.3 Preparation and Use of the Modified PNNP

Scheme 4.3: Preparation of key intermediate **7**.

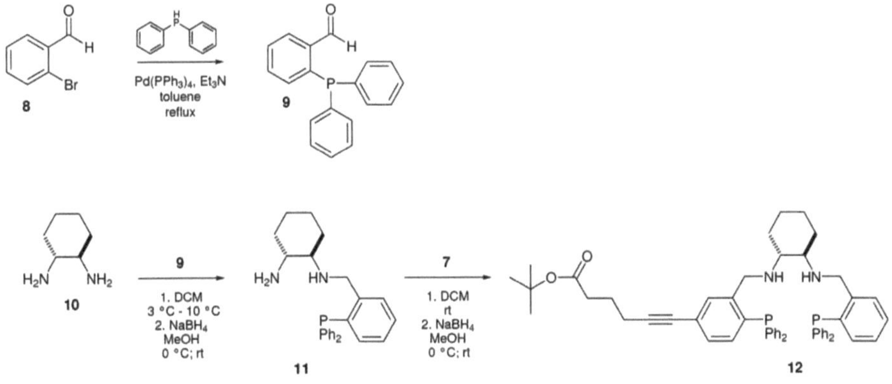

Scheme 4.4: Preparation of key intermediate **12**.

4 Modification of a Ruthenium Catalyst

Using compound **8**, the coupling with diphenylphosphine to yield **9** resulted more efficient than the reaction of **6** under analogous conditions (Scheme 4.4). A new procedure, however, had to be developed for the monosubstitution of (R,R)-diaminocyclohexane (**10**), since the (quite sophisticated) procedure reported by Laue could not be followed successfully [88]. A yield of 72 % of compound **11** was achieved when dichloromethane (DCM) was used as solvent for the first step—monosubstitution of dimaniocyclohexane yielding the intermediate (mono)imine at reduced temperature, in high dilution, and with careful addition of **9**—and methanol for the second step—reduction by treatment with sodium borohydride—in a comparatively simple procedure. The next step, the preparation of the second key intermediate (**12**), was performed at room temperature by adding **7** to a solution of **11** in DCM to form the imine, which again was subsequently reduced by treatment with sodium borohydride in methanol.

Cleavage of the *tert*-butyl ester (**12**) by treatment with trifluoroacetic acid (TFA) yielded the desired "modified PNNP" ligand (**13**), which was transformed to the ruthenium complex (**14**) via reaction with dichlorotetrakis(dimethyl sulphoxide)ruthenium(II) (Scheme 4.5). Thus, 10 synthetic steps were required to prepare the modified Ru-PNNP complex so that it was ready for immobilization. Although the average yield was ∼70 %, the overall yield of **14** was only about 8 % (calculation based on the longest linear sequence of seven steps, thus three further steps are not included).

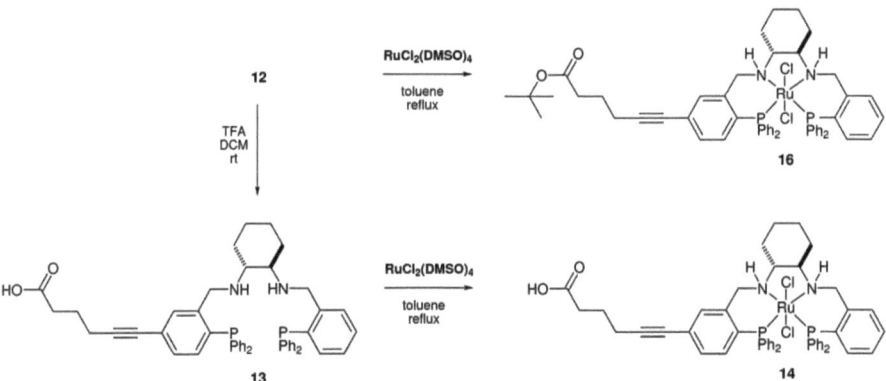

Scheme 4.5: Preparation of the modified Ru-PNNP complex.

Metal insertion into compound **12** instead of acidic cleavage gave the "protected modified Ru-PNNP" complex (**16**), which was used to test the catalytic activity in homogeneous applications, avoiding potential inhibiting effects of the carboxyl group. Reactions performed under equal conditions indeed showed that modified (**16**) and unmodified Ru-PNNP (**18**) performed almost equally as catalysts in the ATH of acp. Thus, the structural modification—the triple bond especially could have been suspected to have an inhibiting effect—had no impact on the catalytic center.

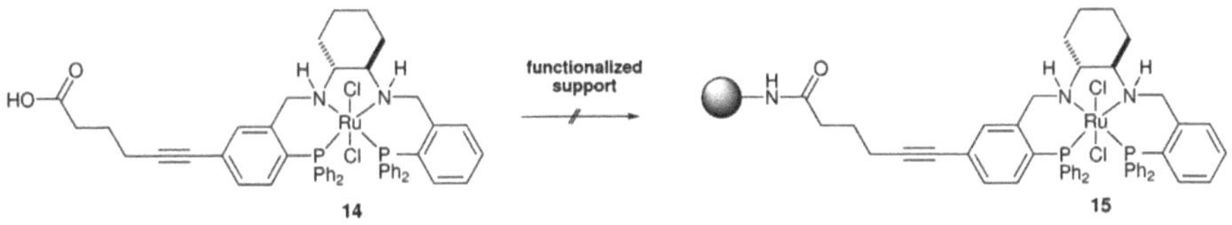

4.3 Preparation and Use of the Modified PNNP

Scheme 4.6: Coupling of modified Ru-PNNP.

4.3.2 Attempts at Immobilization

Due to the more promising results of the ruthenium catalyst, the main focus of the immobilization attempts lay with the generation of a solid polymer-supported Ru-PNNP system. Generally, two pathways could be applied to build the supported catalyst: either coupling of the modified complex **14** (Scheme 4.6) or coupling of the modified ligand **13** (Scheme 4.7) with subsequent metal insertion. While the first approach would allow for better control (unspecific metal–surface interaction could be largely avoided), the latter would provide a more flexible platform, permitting not only the use of ruthenium, but, in principle, also iridium or rhodium either for ATH operations or for other reactions. A limitation of the latter approach, however, derives from the fact that the "pure" ligand is much more sensitive to oxidation than the complex. Hence, coupling and storing have to be performed even more carefully since an attached oxidized species cannot be separated.

The coupling of carboxylic acids with amine compounds is preceded by activation of the carboxyl group. A great number of activation agents have been developed in the context of peptide coupling in organic synthesis, each of which is applicable to specific operations [175]. Coupling to the polymer chips — in this case membrane slices were used — was tested with model reagent 3-(4-methoxyphenyl)propionic acid, activated by o-(benzotriazol-1-yl)-N,N,N',N'-tetramethyluronium tetrafluoroborate (TBTU) in a dimethyl formamide (DMF) solution of diisopropylethylamine (DIPEA) at room temperature. Analysis via infrared (IR) spectroscopy indicated the successful attachment.

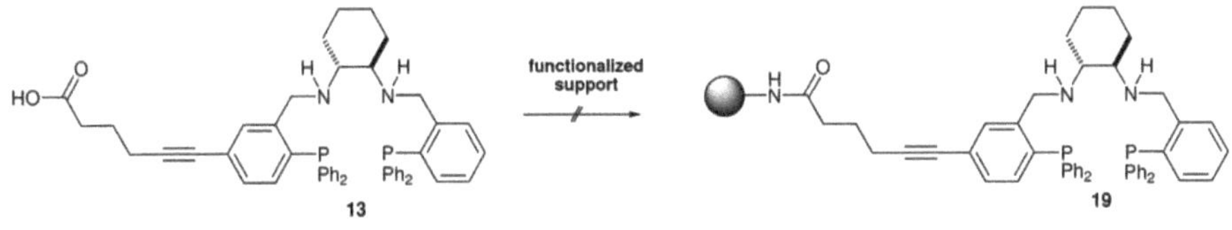

Scheme 4.7: Coupling of modified PNNP.

Thus, initial attempts at the coupling of compound **14** to generate PB and PC-supported versions of the ruthenium complex **15** were performed using TBTU under the same conditions. The supports were stained slightly yellow after the procedure, but a clear indication of a covalent attachment by IR analysis was not obtained, and the supports were not catalytically active. As displayed in Table 4.2, the application of further coupling reagents was tested: 1,1'-carbonyldiimidazole (CDI), N,N'-dicyclohexylcarbodiimide (DCC),

4 Modification of a Ruthenium Catalyst

Table 4.2: Attempts at immobilization to aminated supports.
General conditions: anhydrous and inert, excess of DIPEA over specified amino groups, room temperature, 16–24 h.

Entry	Compound	Support	Amine loading [µmol g^{-1}]	Coupling reagent/ additive/solvent
1	Ru-PNNP	PB	3.8	TBTU/DMF
2	Ru-PNNP	PC	8.6	TBTU/DMF
3	Ru-PNNP	PC	8.6	DCC/NHS/DMF
4	Ru-PNNP	PC	8.6	DCC/HOBt/DMF
5	Ru-PNNP	PC	8.6	DIC/HOAt/DMF
6	PNNP	PB	20	TBTU/DMF
7	PNNP	PC	14	CDI/DMF
8[a]	PNNP	-	-	DCC/NHS/THF

[a] proof of NHS-ester formation attempted

N,N′-diisopropylcarbodiimide (DIC), either with or without an additive such as N-hydroxysuccinimide (NHS), hydroxybenzotriazole (HOBt), or 1-hydroxy-7-azabenzotriazole (HOAt), all without success. Inter-molecular interactions between the carboxylic group and the metal center were considered a possible explanation for the unsuccessful coupling. The second pathway — linkage of the PNNP ligand (13) and subsequent metal insertion — therefore became the focus.

The insertion of ruthenium into the PNNP ligand (both original and modified version) was performed in toluene. The polymer supports, however, were not stable in this solvent (even less under reflux conditions), and another solvent was required to prove that this strategy was, in principle, valid. Using 17 as model ligand, isopropylic alcohol was found to be an effective alternative to toluene, although it has a significantly lower boiling point. Nevertheless, all attempts at linking 13 to one of the supports were unsuccessful; IR analysis did not indicate any attachment, nor did the insertion of ruthenium or the use of these supports in conjunction with the iridium precursor lead to a catalytically active system. At this point, it was assumed that the failure of coupling was due to unsuitable procedures. This could not be definitively determined, however, and finally the search for more appropriate coupling reagents and conditions was abandoned.

4.3.3 Concluding Remarks

A modified version of the Ru-PNNP catalyst was prepared via a new 10-step convergent synthesis route. Although many attempts were made, the attachment to surface-functionalized polymer supports via amide formation reaction failed. A significant improvement in the overall yield of less than ∼8 % in the preparation of compound 14 was not achieved, and upscaling was considered incalculable, so only small amounts could be produced. Even if an appropriate coupling procedure had been developed, the low overall yield was considered a major hindrance for the detailed study of the catalyst properties and optimization within a reasonable time and cost frame. Thus, taking into account the aim of this work, the decision was made to focus on the use of the second catalyst system which represents the latest developments in the field (Chapter 5). Further considerations on the efficiency of immobilized (PNNP-based) catalysts are presented in Chapter 6.

5 Immobilization of a Rhodium Catalyst

The search for another complex which was suitable for immobilization and use as catalyst in the ATH of prochiral substrates, resulted in the tethered tetramethyl cyclopentadienyl rhodium(III) p-toluenesulfonyl-1,2-diphenylethylenediamine complex (henceforth *tethered Rh-TsDPEN*, Fig. 5.1, cf. section 2.4.3) which was presented by Wills et al. in 2005 [95]. It was shown to be a highly enantioselective, active, and versatile catalyst for the transfer hydrogenation of various ketones in TEAF, and from reports on the use of analogous catalysts, an excellent performance in aqueous media was anticipated [176, 177]. The special feature of the ligand is the connection (tether) between the coordinating η^5-cyclopentadienyl moiety and the chiral backbone, which contains two electron-donating nitrogen atoms. The resulting multidenticity provides a high complex stability. Moreover, tethered versions of TsDPEN-based catalysts were demonstrated to have significant advantages in terms of reaction rate and versatility over the untethered variants for various substrates [178].

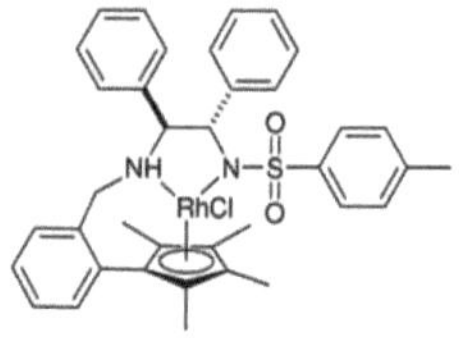

Figure 5.1: Wills´s "tethered Rh-TsDPEN" catalyst.

5.1 Synthetic Modification

5.1.1 Ligand Preparation

Analogous to the synthetic modification of the PNNP ligand (cf. Chapter 4), the introduction of hex-5-ynoic acid *tert*-butyl ester as the protected spacer/linker moiety via Sonogashira coupling reaction was chosen to create a modified version of the tethered TsDPEN ligand, ready for attachment to aminated supports. For this a halide functionality at any site of the ligand was required. Since the 4-position of the phenylsulfonyl moiety was favored for the introduction, toluenesulfonyl was substituted for bromobenzenesulfonyl. Two strategies for the preparation of the ligand were employed, differing in the order of the synthetic steps. One consisted in first coupling the diamine with bromobenzene-sulfonyl chloride to obtain the "bromo-functionalized TsDPEN" (Scheme 5.1 **b**), which afterwards was coupled with hex-5-ynoic acid tert-butylester via palladium-catalyzed Sonogashira cross coupling (Sch. 5.1 **c**).

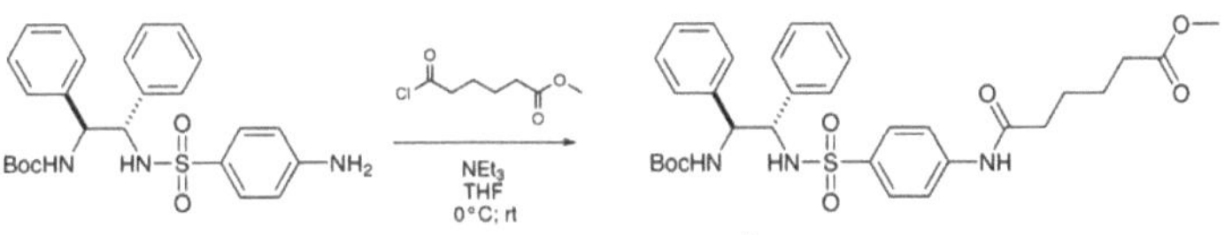

5.1 Synthetic Modification

Scheme 5.3: Introduction of methyl adipoyl chloride.

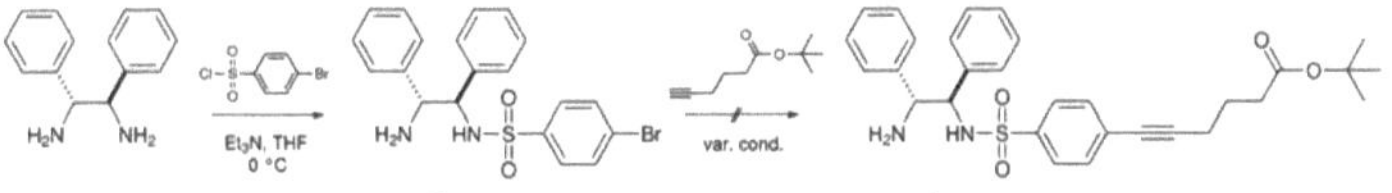

Scheme 5.1: Introduction of hex-5-ynoic acid-tert-butylester — first Attempt.

Unfortunately, this approach did not lead to the desired product. The "bromo-functionalized TsDPEN" (Sch. 5.1 b) was obtained in high yields, but cross coupling with the alkynyl in the second step was not achieved. This was ascribed to coordination of the diamine to the palladium catalyst, which led to inhibition of the latter. The reversed order, first the cross coupling of bromobenzene-sulfonyl chloride with hex-5-ynoic acid tert-butyl ester and then the reaction with the chiral diamine, failed in the very first step (Scheme 5.2).

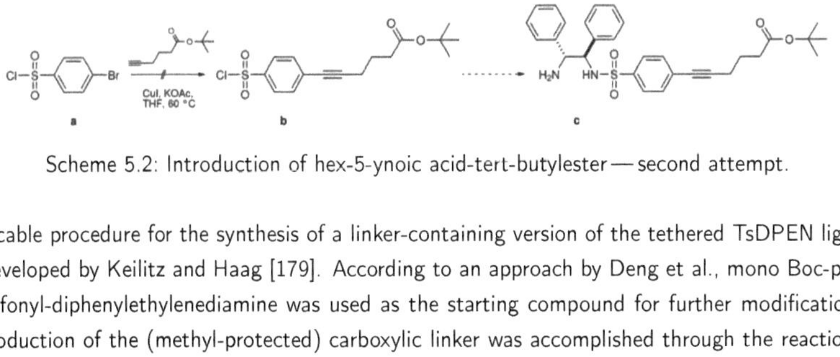

Scheme 5.2: Introduction of hex-5-ynoic acid-tert-butylester — second attempt.

An applicable procedure for the synthesis of a linker-containing version of the tethered TsDPEN ligand was finally developed by Keilitz and Haag [179]. According to an approach by Deng et al., mono Boc-protected aminosulfonyl-diphenylethylenediamine was used as the starting compound for further modification [180]. The introduction of the (methyl-protected) carboxylic linker was accomplished through the reaction of the "amino-functionalized TsDPEN" with methyl adipoyl chloride (Scheme 5.3). In four further steps, the desired linker-containing ligand (**20**, Scheme 5.4) was obtained in ~40 % overall yield (calculated on the basis of the seven linear steps of the convergent route).

5.1.2 Catalyst Formation

The insertion of rhodium into the ligand could either be performed before the ligand was coupled to the support (Scheme 5.4, path **A**) or afterwards (path **B**). In order to keep the number of synthetic steps on the solid phase as small as possible and to avoid unspecific binding of rhodium onto the surface, the coupling of the whole complex (path **A**) was favored. Nevertheless, the rhodium insertion could not be performed in methanol, as was done with the unmodified ligand by Wills and co-workers [95], because of the resulting esterification of the linker. Aprotic polar solvents were tested to carry out this reaction, and tetrahydrofuran (THF) proved the most convenient. Methanol, however, could be used as solvent for the insertion in path **B**, or when the preparation of the methyl-protected variant (**23**) was sought, e. g., for homogeneous applications of the catalyst. The resulting NMR spectra of **21** were more complex than those of the methyl-protected version. Characterization of the desired substance was therefore performed via high resolution mass spectrometry and infrared spectroscopy (see Section 7.2.2).

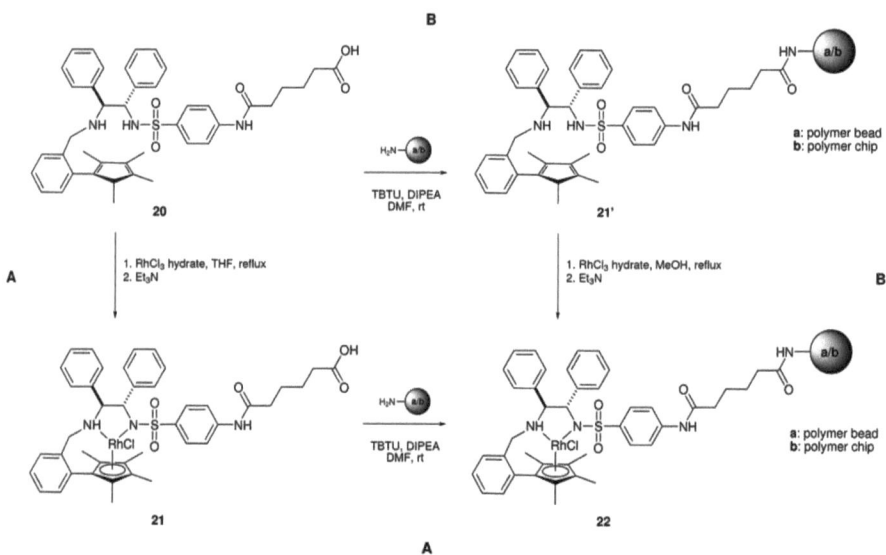

Scheme 5.4: Preparation of the supported modified tethered Rh-TsDPEN catalyst (**22**).

After the successful preparation of compound **21**, the carboxylic linker was activated with TBTU in DMF at room temperature, according to standard amide coupling procedures. The supports had to be prepared via acidic cleavage of the Boc protecting groups in HCl/2-propanol. Thus, diisopropylethylamine (DIPEA) was added as a base to dissolve the hydrochlorides generated. Activated **21** dissolved in DIPEA/DMF was shaken together with the deprotected supports for 20 hours, washed with organic solvents, and dried under reduced pressure. Scheme 5.4 depicts the preparation of the *S,S*-version of the immobilized catalyst. The same procedure applying the antipodal diamine yielded the *R,R*-version; the *S,S*-version, however, was prepared in larger scale and was hence used exclusively in this study.

5.1 Synthetic Modification

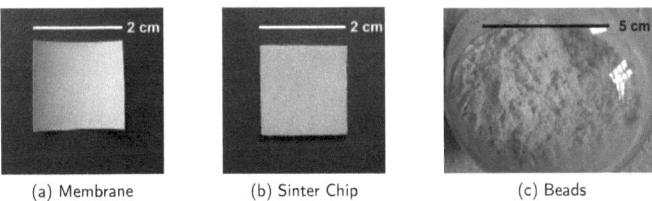

(a) Membrane   (b) Sinter Chip   (c) Beads

Figure 5.2: Supported Catalysts.

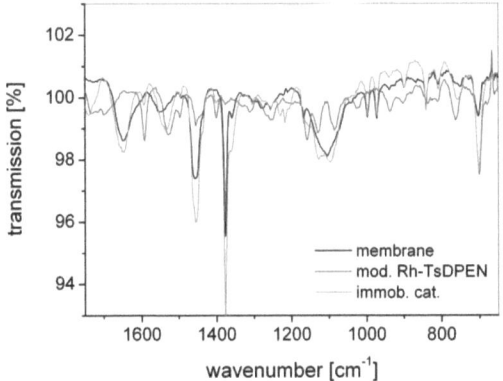

Figure 5.3: IR spectra as proof of coupling.

Successful coupling was indicated by the bright orange color of the supports after the procedure was finished (Figure 5.2). Additionally, IR analysis of the catalyst-loaded PCs yielded phenyl and sulfonamide signals at 1593/1496 cm^{-1} and (1375/)1157 cm^{-1}, respectively (Figure 5.3). The amount of catalyst coupled to the surface was considered consistent with the amount of rhodium which was quantified via ICP analysis (cf. Section 3.1.2). Samples were therefore prepared via microwave digestion in aqueous HNO_3/HCl solution. It was observed that the amount of DIPEA added during the coupling procedure had a significant influence on the amount of catalyst attached to the surface. Based on the number of amino groups determined, four to six equivalents of DIPEA provided a catalyst loading that was more than twice as high as with two equivalents.

5.2 Catalytic Testing

5.2.1 General Remarks

Both PC-supported versions (membrane slices and sinter chips) as well as the PB-supported version of the modified tethered Rh-TsDPEN catalyst were used in ATH reactions. Generally, tests to elucidate basal properties such as substrate scope or reuseability were performed with PCs, since the handling was easier and an ordinary reaction set-up consisting of a standard flask with a magnetic stirrer could be used. Hence, PCs could even be applied to parallel experiments. The PCs were simply added to the reaction mixture and taken out for washing and reuse using forceps without any loss of support material (Figure 5.4 a). Because of the higher mechanical stability, the sinter chip version was favored over membranes, particularly for testing the reusability. However, several drawbacks of the 2-dimensional supports made the application of beads necessary:

- The functionalization of the support materials as well as the immobilization of the catalyst was performed batchwise. As a result, all batches differed slightly in amine and catalyst loading. The advantage of beads was that a major blend, providing a large amount of highly reproducible samples, could be made from various batches.

- A related problem was the catalyst metering, which was less precise when using PCs, even when they stemmed from one batch. Usually slices of equal geometric surface, e.g., 1×1 cm^2, were applied for one series experiment, but an equal amount of immobilized catalyst was not necessarily provided, neither by this method nor by weighing the PCs, due to the deviations of the specific surface. This problem, however, was less severe since deviations were found to be small.

- Nevertheless, reproducible reaction conditions were a problem, since both PBs and PCs have a low density. Through vigorous mechanical stirring, the beads were homogeneously suspended in aqueous solutions, whereas the chips could only be used with magnetic stirring, and floated on the water surface. A more precise adjustment of the stirring speed and the achievement of a homogeneous suspension, which guaranteed highly reproducible conditions, was only possible when beads were applied.

ATH experiments with the PB-supported catalyst (**22 a**) were performed in a jacketed 100-mL flask equipped with a mechanical stirrer, a thermostat, an insight temperature sensor, and an optional argon sweep gas installation (Figure 5.4 b). As depicted in Scheme 5.5, the focus of all experiments performed with the supported tethered Rh-TsDPEN catalysts lay on the enantioselective reduction of prochiral ketones to chiral alcohols and the use of water as a potentially "green" reaction medium [71, 72]. The ATH of acetophenone performed in an aqueous solution of sodium formate at 40 °C was used as the benchmark reaction for the testing of the PC and PB-supported versions of the catalyst.

In contrast to the PNNP-based catalyst, by applying the *S,S*-version of the tethered Rh-TsDPEN the *S*-enantiomer of the product was formed in excess. Both enantiomeric excess and conversion were determined via GC analysis. For the determination of the conversion, samples were taken from the reaction mixture with a syringe, beads (when used) were filtered off, and the samples were diluted in ethanol and injected

5.2 Catalytic Testing

(a) PC-supported catalyst.

(b) PB-supported catalyst.

Figure 5.4: Use of supported catalysts.

into the gas chromatograph. Usually, the *ee* was only determined after the reaction was finished. Two apparatus equipped with different chiral columns were used for this analysis, and slightly different *ee* values were obtained, varying between 98 % and 99 %.

Scheme 5.5: ATH of aryl ketones in an aqueous medium.

5.2.2 Initial Experiments

5.2.2.1 ATH of Acetophenone

In order to guarantee complete wetting of the supports without exceeding the usual range of the substrate to catalyst ratio (S/C), a relatively high dilution was required in all experiments. The PC-supported catalyst (**22 b**) was used at a substrate concentration of $0.1\ \text{mol L}^{-1}$ with an S/C between 200 and 450. Under these conditions, **22 b** provided a higher conversion than both the unsupported analog (**23**) and a soluble hyperbranched PEG-supported version within the same time frame (Table 5.1) [179]. Under the standard conditions for the use of catalyst beads (see Section 5.3), however, no comparison between the PB and the non-supported catalyst was possible. Apparently, the substrate concentration of $0.04\ \text{mol L}^{-1}$ was too low, and after a time the catalyst tended to adhere to the stirrer tool. These results are remarkable because a significant limitation of the rate due to mass diffusion effects should have been expected from a water/oil/solid triphase reaction. However the opposite effect, a rate enhancement, was observed when the immobilized catalyst was used. This acceleration is presumably derived from an accumulation effect of the support material which causes higher concentrations of the reactants in the microenvironment of the catalyst (cf. Section 2.4.6.2).

Table 5.1: Performance of different versions of the modified tethered Rh-TsDPEN.
Reaction conditions: acp 1 mmol, HCO$_2$Na 5 mmol, water 10 mL, S/C ≈ 430, 40 °C.

Entry	Catalyst	Time [h]	Conversion [%]	ee[a] [%]
1	soluble PEG-supported	4	18	98
2	PC-supported (22 b)	4	57	99
3	non-supported (23)	4	37	98

[a] excess of the S-enantiomer

Generally, the results obtained with the PC-supported catalyst in organic media were far inferior to those obtained in water. In pure TEAF, the supports were tinged black, and the catalyst was almost inactive. When TEAF was diluted with DCM, a somewhat higher conversion was achieved, but it was still not in the range of that obtained with the sodium formate/water system; under otherwise equal conditions, the TEAF/DCM medium provided a conversion of <10 % in six hours whereas in HCO$_2$Na/H$_2$O almost full conversion was achieved. The reaction proceeded even more slowly in pure 2-propanol. These results may support the above-mentioned suggestion of a potential accumulation effect. If the substrate (at a low concentration) is completely dissolved in the organic solvent and not accumulated in the environment of the catalyst, the reaction may become diffusion-controlled. The addition of methanol as a co-solvent to the sodium formate/water system, however, slightly enhanced the rate and did not affect the enantioselectivity.

In contrast to the Ru-PNNP catalyst, no induction phase was observed with the immobilized tethered Rh-TsDPEN catalyst in aqueous media. This may indicate that the formation of the active species proceeds rapidly compared to the rate-determining step, i. e., the hydrogen transfer [112]. When the bead-supported catalyst (22 a) was applied under the standard conditions, high reproducibility of the conversion over time was achieved. In four independent experiments, the maximum standard deviation was about 3 % (Figure 5.5). Other conditions, however, provided less reproducible results (see below). A significant difference between reactions performed in air and in an argon atmosphere (closed flasks without gas disengagement in either case) was not observed.

5.2.2.2 Recycling

Catalyst recycling was performed either with or without washing the supports with water and/or organic solvents after each run. A high decrease in activity was observed within the first three cycles when the supports were repeatedly used in the standard sodium formate/water system without washing or further treatment (Table 5.2). This decrease is not fully explained by the metal leaching, which was determined via ICP-MS analysis. A leaching of 56 ppm (rhodium per total amount of substrate), which corresponds to approximately 2.5 % of the initial rhodium content, was observed after the first run. After two runs, approximately 4 % of the rhodium had leached into the liquid phase, but the conversion dropped by nearly 50 % in the third run (cf. entry 1 and 3, Table 5.2). In further runs, the amount of rhodium loss was significantly lower. Thus, unspecifically bound 21 or rhodium is assumed to have washed away during the first runs. The ee values were determined to be 98 % in all runs, and thus proved to be independent of the

5.2 Catalytic Testing

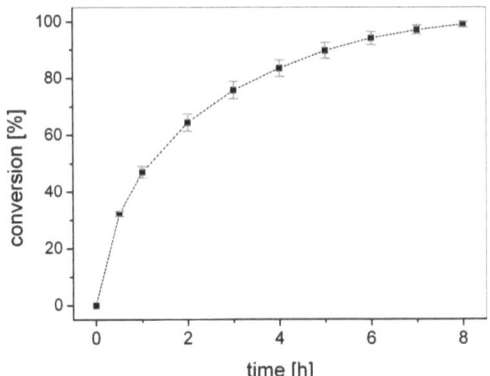

Figure 5.5: Reproducibility of the standard experiment.
Reaction conditions: acp 0.04 M, HCO$_2$Na 0.4 M, total volume of water/HCO$_2$Na/acp 100 mL, PB-supported catalyst 2 g ($\approx$ 0.113 mM Rh), 40 °C, 1 200 rpm, argon overflow.

conversion in the investigated range.

Table 5.3: Recycling of PC-Supported catalyst with intermediate washing.
Reaction conditions: acp 1 mmol, HCO$_2$Na 5 mmol, water 10 mL, PC-supported catalyst, S/C $\approx$ 430, 40 °C.

Run	Time [h]	Conversion [%]	ee[a] [%]
1	4	57	99
2	4	51	98
3	4	44	99
4	4	43	99
5	16	79	99
6	4	27	98
7	4	23	98

[a] excess of the S-enantiomer

A decrease in catalytic activity was also observed in recycling experiments in aqueous sodium formate with intermediate washing and drying (Table 5.3). In a reaction that was run for four hours to a conversion of 57 % in the first run, a drop of 6 % and 7 % (referred to 100 % conversion) was observed in the second and third run, respectively. After a fourth run with almost identical conversion, the fifth run was performed overnight, achieving 79 % conversion within 16 hours. The subsequent sixth run only achieved 27 % conversion within four hours. Again, ee values were stable throughout the experiment. From these results, it was concluded that washing the supports did not have a significant effect on the catalyst reusability. High losses of activity

Table 5.2: Recycling of PC-supported catalyst without intermediate washing.
Reaction conditions: acp 1 mmol, HCO$_2$Na 5 mmol, water 10 mL, PC-supported catalyst, S/C $\approx$ 430, 40 °C.

Run	Time [h]	Conversion [%]	ee[a] [%]	Leaching [ppm][b]	[%][c]
1	8	100	98	56	2.5
2	16	100	98	35	1.6
3	8	53	98	14	0.6
4	16	68	98	n. d.	n. d.
5	8	55	98	n. d.	n. d.
6	16	59	98	12	0.6

[a] excess of the S-enantiomer
[b] detected amount of rhodium per total amount of substrate, n. d. = not determined
[c] detected amount of rhodium per initial amount of catalyst, n. d. = not determined

were observed in either case when the reaction was run to high conversion. Thus, a catalyst deactivation due to byproducts of some kind was considered a possible explanation and investigated in further experiments (see Sections 5.2.3 and 5.2.4). The use of argon as protective gas in closed flasks could again not improve the performance.

5.2.2.3 Substrate Scope

The versatility of the catalyst was tested in ATH reactions with different aryl ketones as substrates. High values of the enantiomeric excess were achieved in all cases, ranging from 87 % to 98 % (Table 5.4). In contrast, the conversion within six hours reaction time differed more strongly, presumably due to different solubilities in water and the functional matrix of the support material. This was definitely observed with 4'-nitroacetophenone (Entry 2), a solid with low solubility in water.

Table 5.4: Scope of tested substrates.
Reaction conditions: substrate 1 mmol, HCO$_2$Na 5 mmol, water 10 mL, PC-supported catalyst, S/C $\approx$ 430, 40 °C.

Entry	Indication	Substrate R^1	R^2	Time [h]	Conversion [%]	ee[a] [%]
1	acp	H	H	6	97	98
2	-	4'-NO$_2$	H	6	36	87
3	-	4'-Cl	H	6	71	95
4	-	4'-OMe	H	6	79	97
5	-	H	Me	6	58	98

[a] excess of the S-enantiomer

5.2.3 Effect of Temperature and Atmosphere

For the determination of temperature and atmosphere effects on the ATH of acetophenone, the PB-supported catalyst was used. A normal correlation of temperature and conversion rate was observed in experiments conducted between 10 °C and 40 °C in the closed 100-mL flask without applying any protective gas or sweep gas. Hence, within this range an increase in temperature effected an increase in rate. At 50 °C the reaction started faster than at 40 °C but was slower at the end; at 60 °C a significantly lower initial rate was observed, and the reaction leveled off at about 40 % conversion (Figure 5.6 a). The retarding effect on the conversion rate at temperatures above 40 °C presumably derives either from inhibition due to interaction with byproducts which emerge only at higher temperatures, or from catalyst decomposition.

Reuse experiments indicated that decomposition was less probable, because the employment of catalyst beads taken from two experiments, one conducted at 40 °C, the other conducted at 60 °C, gave similar results in terms of conversion and *ee* in the corresponding runs at 40 °C after recycling. Nevertheless, the rate was significantly lower than in the first run at 40 °C. Two effects that decreased the catalyst activity had to be distinguished:

- a temperature-related effect which was reversible by washing the supports, and

- another effect which was not reversible (at least not by washing the supports, see above).

However, it is known that CO_2 emerges as a byproduct during the reaction, and a CO_2 insertion [181] as well as the inhibiting effect of CO_2 headspace gas in ATH reactions performed in aqueous solutions of sodium formate [112] have been reported. Furthermore, the generation of CO from formic acid in ATH reactions has been noted [116, 182].

In order to remove the emerging gases, an argon overflow was applied. As a result, the conversion rate at 60 °C was significantly higher than without gas disengagement, and full conversion was successfully achieved (Figure 5.6 a). Nevertheless, apart from a trend toward higher reaction rates at temperatures between 50 °C and 70 °C when employing the sweep gas, fully consistent and reproducible data could not be generated. This was ascribed to unstable and insufficient gas disengagement, and therefore a gas entrainment impeller was used for more efficient dispersion of argon into the reaction mixture. Unfortunately, the catalyst beads could not be suspended homogeneously under these conditions because the buoyancy was too strong. Reliable data in experiments above 40 °C could therefore not be obtained.

For a more detailed study of the generation of carbon monoxide and carbon dioxide, samples taken from the gas phase in the flask during a reaction at 60 °C were analyzed. A significant amount of CO could not be detected; only the generation of CO_2 from sodium formate was proven by GC analysis even when no substrate was present. The impact of carbon dioxide on the reaction was investigated by comparing the conversion in experiments conducted at 40 °C with argon overflow, in a closed flask, and in a CO_2 atmosphere (CO_2 sweep gas), respectively. Surprisingly, an increasing amount of CO_2 in the reactor effected an increase in reactivity rather than inhibiting the reaction (Figure 5.6 b). This was presumed to derive from the reduced solution basicity due to solubilized carbon dioxide, and in fact pH values at the end of the reactions were

5 Immobilization of a Rhodium Catalyst

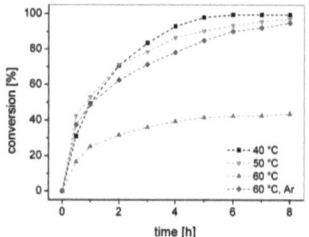
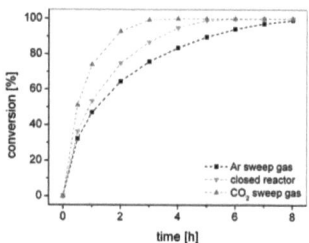

(a) Temperature variation.
Reaction conditions: acp 0.04 M, HCO$_2$Na 0.4 M, PB-supported catalyst 1.5 g ($\sim$0.105 mM Rh.), 1 200 rpm

(b) Atmosphere variation.
Reaction conditions: acp 0.04 M, HCO$_2$Na 0.4 M, PB-supported catalyst 2 g ($\sim$0.113 mM Rh), 40 °C, 1 200 rpm.

Figure 5.6: Variation of temperature and atmosphere within the flask.

significantly lower when no gas disengagement was carried out (closed reactor: pH 8.6) or when CO$_2$ was used as sweep gas (pH 7.1), compared to experiments with argon overflow (pH 9.4).

Although it could not be confirmed via GC analysis, it is most likely that the inhibition at temperatures above 40 °C is due to the generation of CO. The carbon monoxide is believed to coordinate quickly to the metal center, inhibiting the catalytic activity and hence the production of additional CO. Blacker and Thompson, who studied the use of a (non-tethered) Rh-TsDPEN catalyst in the TEAF system on a larger scale, stated that the amount of carbon monoxide produced was small and that coordination to the catalyst was reversible [116, 183]. This would confirm the above findings and assumptions. However, a temperature dependency of the assumed production of CO has not been mentioned in the literature (to the best of the authors' knowledge); only the need for carbon dioxide removal, which is apparently not required when 22 is used as catalyst, has been stated [116].

5.2.4 pH Dependency of Activity, Enantioselectivity, and Reusability

Both enantioselectivity and activity in ATH reactions catalyzed by transition metal complexes based on TsDPEN or its derivatives have been shown to be pH-dependent [184, 185]; a few studies have been conducted to elucidate the reasons for this [112, 183]. In order to investigate the influence of pH on the catalyst performance and to determine the optimal reaction conditions, the initial solution pH was adjusted to values between 1 and 12 by applying sodium formate, formic acid, and mixtures of both, or adding NaOH(aq) or HCl(aq) to further increase or decrease the pH, respectively (see Table 5.5). For these tests, the sinter chip immobilized version of the catalyst was used due to its easier handling compared to the micro particles. In almost all experiments the pH increased with increasing conversion. Only when the initial pH was adjusted to 9.5 by adding a small amount of NaOH ($\sim$1 µmol) to the sodium formate/water solution was a decrease of the pH to a value of 8.5 observed during the reaction (Table 5.5, Entry 9).

The optimal initial pH in terms of activity was found to be between 3 and 4, giving values of about 4 to 5 at the end of the reactions (Entries 3 to 5). By performing analogous experiments following the more

5.2 Catalytic Testing

Table 5.5: ATH of acetophenone under different pH conditions.
Reaction conditions: acp 1 mmol, HCO$_2$Na or HCO$_2$H/HCO$_2$Na 10 mmol, water 10 mL, PC-supported catalyst, S/C ≈ 200, 40 °C.

| Entry | HCO$_2$C/HCO$_2$Na [mmol] | pH | 2 h | | pH | 4 h | | Leaching[c] [%] |
			Conv. [%]	TOF[a] [h^{-1}]		Conv. [%]	ee[b] [%]	
1	(+ HCl)10/0	1.0	0	0	1.1	3	11	2.4
2	10/0	1.7	61	62	2.1	83	75	n. d.
3	7/3	3.1	89	86	3.7	>99	97	n. d.
4	5/5	3.5	91	88	4.3	>99	98	n. d.
5	3/7	3.9	94	94	5.1	>99	98	1.9
6	1/9	4.5	81	80	7.0	>99	98	n. d.
7	0.03/9.97	5.6	67	68	8.2	93	98	n. d.
8	0/10	7.5	61	60	8.4	81	97	2.2
9	0/10 (+ NaOH)	9.5	59	56	8.5	79	98	n. d.
10	0/10 (+ NaOH)	12.3	8	9	12.3	10	90	1.8
11[d]	5/5	3.5	92	99	4.4	>99	98	n. d.
12[e]	5/5	3.5	61	64	4.1	92	97	n. d.

[a] For the calculation of TOF values, the concentration of applied catalyst was determined from the mass of each sinter slice.
[b] excess of the S-enantiomer
[c] detected amount of rhodium per initial amount of catalyst, n. d. = not determined
[d] using catalyst chip from entry 1
[e] using catalyst chip from entry 10

sophisticated approach of applying the PB-supported version of the catalyst, it was confirmed that 3.9 was the optimal pH value rather than 3.5. Maximal ees (98 %) were achieved at initial pH values between 3.5 and 5.6 (Entries 4 to 7).

The reaction was extremely slow at both very low and very high pH levels, and also lower ees were obtained. In order to investigate a potential deactivation effect from strongly acidic and strongly basic media, the catalyst chips which had been used under those conditions (Entry 1, pH 1 and Entry 10, pH 12.3, respectively) were recovered and applied in moderately acidic HCO$_2$H/HCO$_2$Na/H$_2$O mixtures (Entries 11 and 12, respectively, pH 3.5). It was found that the strongly acidic conditions, which inhibited the conversion in the first run, had no significant influence on the catalyst performance in the second run when the pH was adjusted to 3.5; values of conversion and ee were in agreement with those obtained with the non-pretreated catalyst or even somewhat better (Entry 11 cf. Entry 4). The catalyst chip which had been used in the strongly basic HCO$_2$Na/NaOH/H$_2$O mixture (pH 12.3) also showed an increased activity at a pH of 3.5, but it lagged behind that of the non-pretreated catalyst (Entry 12 cf. Entry 4). This indicates that the inhibition under acidic conditions is completely reversible, whereas in basic media it is only partly reversible.

Increased metal leaching under basic conditions was considered a possible explanation for these results. Thus, the rhodium content of the solutions of four experiments performed under different pH conditions was analyzed via ICP-OES analysis. The amount of rhodium in the solutions was found to be between 1.8 % and 2.4 % of the initial amount on the catalyst chips. Since a correlation between leaching and solution pH

5 Immobilization of a Rhodium Catalyst

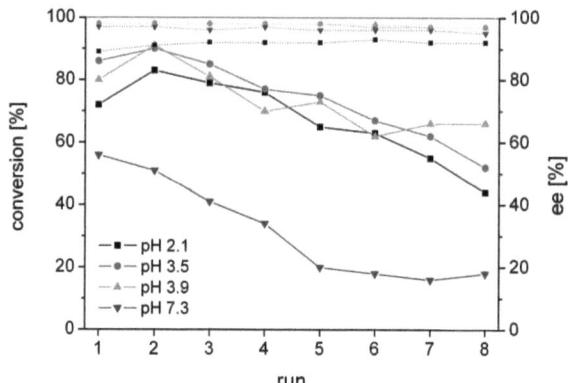

Figure 5.7: Multiple catalyst reuse under different pH conditions. Reaction conditions: acp 0.1 M, HCO_2Na or HCO_2H/HCO_2Na 0.5 M, water 2.5 mL, PC-supported catalyst, S/C ≈ 200, 40 °C, reaction time 2 h per run.

was not evident (cf. Entries 1, 5, 8, and 10), the reduced reactivity under basic conditions is more likely to derive from an inactive catalyst species [186]. The increase of the pH value, likely due to the generation of hydroxide ions from formate, is accompanied by an increasing conversion of acp. An irreversible or only partly reversible inhibiting effect from OH^- ions is presumed to explain both the reduced reusability when the reaction was run to high conversion in HCO_2Na/H_2O (cf. Section 5.2.2.2) and the low reactivity under strongly basic conditions.

The recyclability was further tested under different pH conditions. The data points shown in Figure 5.7 represent conversion (full lines) and ee (dotted lines) of three series of experiments, each with eight consecutive runs performed under equal conditions. The initial pH values were adjusted to 2.1, 3.9, and 7.4. After two hours reaction time, the catalyst chips were taken out, washed, dried, and used for the next run, while conversion and ee were determined via GC-MS analysis. The best performance in terms of recyclability was achieved at an initial pH of 3.9, providing a TTN of 1 255 over eight runs, with the catalyst still active. This good reusability under moderately acidic conditions was accompanied by the highest conversion and enantioselectivity (constant ~98 % ee). The total rhodium leaching over eight runs was determined to be around 5 %. In the acidic mixtures, the conversion increased from the first to the second run and declined in the subsequent runs. This increase was ascribed to the basic conditions of the coupling procedure which may have affected the catalyst performance in the first run.

Taking into account the results obtained from varying the temperature and atmosphere and the increase in pH observed during the course of the reaction, a more complete picture of the process can be drawn (Scheme 5.6). Thus, besides the reduction of acetophenone to phenylethanol, in which simultaneously carbon dioxide and hydroxide are generated from formate and water, there are probably two side reactions

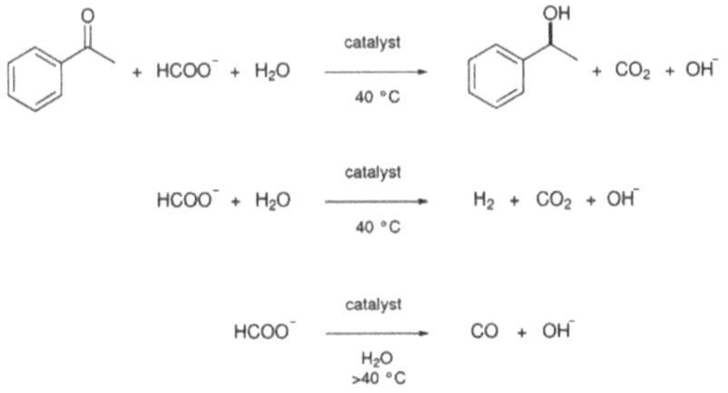

Scheme 5.6: ATH of aryl ketones in formate/water.

occurring. One is the decomposition of formate in the presence of water to molecular hydrogen and carbon dioxide, and the other is decarbonylation, which is assumed to occur only at temperatures above 40 °C when 22 is used as the catalyst. These side reactions can only indirectly diminish the yield of the desired product (catalyst poisoning by carbon monoxide, as discussed), since the aryl ketone substrate is not involved.

5.2.5 Effect of the Concentrations of Acetophenone and Sodium Formate

The linear relationship between rate and substrate concentration was shown to be valid up to about 0.1 mol L^{-1} of acp when 1.5 g of the PB-supported catalyst were applied (Figure 5.8 a). In this concentration range the beads were homogeneously or quasi-homogeneously suspended (Figure 5.9 a). At higher substrate concentrations, though, the reaction rate decreased due to bead agglomeration within the emulsified acp. The size of the agglomerates increased with increasing concentration and decreasing rate. At an acp concentration of about 0.14 mol L^{-1} firm "balls" of about 0.75 cm in diameter were formed, and almost no conversion occurred (Figure 5.9 b). The inhibition due to aggregation impeded the investigation of a substrate inhibition effect on the molecular level, which has been observed with the unsupported variant at acp concentrations above 1.3 mol L^{-1} [112].

Figure 5.8 b shows the conversion of acp over time at different hydrogen donor to substrate ratios. It is apparent that the acceleration of the reaction due to an increasing excess of sodium formate follows a saturation curve. Thus, the enhancement of the reaction rate brought about by the doubling of the amount of sodium formate—from a hydrogen donor to substrate ratio of 5 to a ratio of 10—is rather small. Ratios above 10 were therefore not investigated.

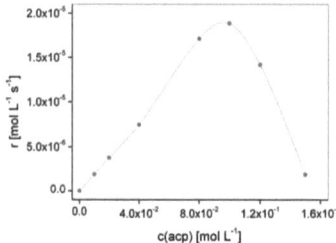
(a) Variation of the concentration of acp.
Reaction conditions: constant 10/1 ratio of HCO_2Na to acp, PB-supported catalyst 1.5 g (∼0.105 mM Rh), 40 °C, 1 200 rpm.

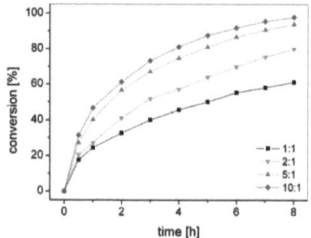
(b) Variation of the concentration of HCO_2Na.
Reaction conditions: constant substrate concentration 0.04 M, PB-supported catalyst 2 g (∼0.113 mM Rh), 40 °C, 1 200 rpm, argon overflow.

Figure 5.8: Impact of the concentrations of substrate and hydrogen donor.

(a) Even suspension at lower concentrations.

(b) Aggregation of beads at higher concentrations.

Figure 5.9: Influence of the concentration of acetophenone on the suspension of beads.

5.3 Kinetic and Mechanistic Investigations

A systematic series of experiments were performed in order to determine the kinetic parameters, taking into account the above-mentioned results that reveal the rate of the ATH reaction under different conditions. As introduced in Section 2.4.4, intensive experimental and theoretical investigations of the asymmetric transfer hydrogenation of acetophenone in aqueous media were reported by Liu, Xiao et al. [112]. A first-order rate dependence with respect to the substrate as well as to the catalyst concentration was proposed and experimentally confirmed for an acp concentration of up to 1.3 mol L^{-1} and excess sodium formate using a non-tethered η^6-arene ruthenium(II)-p-toluenesulfonyl-1,2-diphenylethylenediamine complex in one-phase water/DMF solution [74, 112]. This rate dependence (Equation 2.4.2) served as the starting point for the present study: the second-order model was applied to fit the experimentally determined concentration of acp over time. All experiments were exclusively performed using the bead-supported version of the catalyst (22 a).

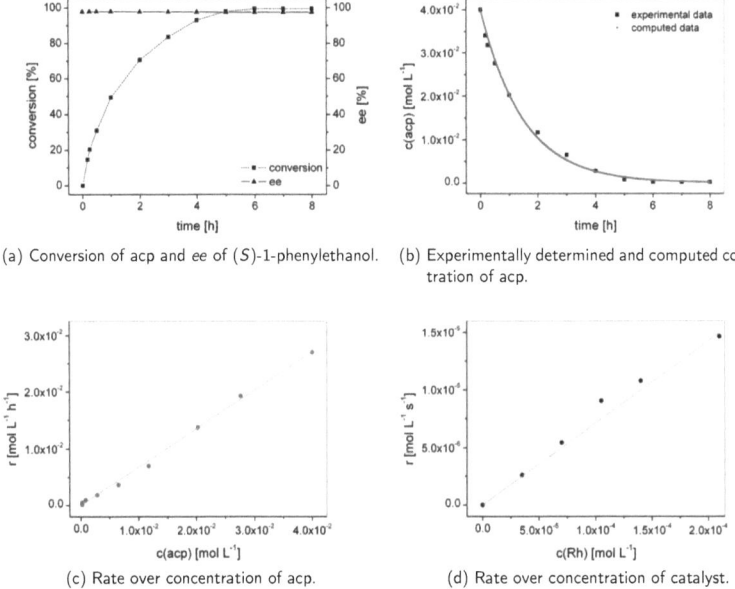

Figure 5.10: ATH of acp under moderately basic (standard) conditions.
Reaction conditions: acp 0.04 M, HCO$_2$Na 0.4 M, PB-supported catalyst 1.5 g (~0.105 mM Rh), 40 °C, 1 200 rpm, total reaction volume 100 mL.

5.3.1 Basic Reaction Conditions

In the standard experiment, 1.5 g of the PB-supported catalyst with a rhodium content of 7 μmol g^{-1} was used, and a tenfold excess of sodium formate (0.4 M) over acetophenone (0.04 M) in neat water was applied. The flask was kept closed except when samples were taken, and no gas disengagement or protective gas was used. A conversion of more than 99 % with an enantiomeric excess of 98 % was achieved after six hours (Figure 5.10 a). The solution pH increased during the reaction from about 7.8 to 8.6. Kinetic and activation parameters were calculated from averaged results of two experiments performed under equal conditions.

Despite the apparent differences between **22 a** and the catalytic system used by Liu and co-workers [112], the model provided good agreement with the experimental results (Figure 5.10 b). Accordingly, a linear dependence of the reaction rate on the acp concentration (Figure 5.10 c) as well as a linear dependence of the reaction rate on the amount of catalyst (Figure 5.10 d) were found in the investigated range. From these experiments, a value of 1.9 ± 0.2 L mol^{-1} s^{-1} was calculated for the rate constant k.

The activation parameters were calculated from experiments conducted between 10 °C and 40 °C, since at higher temperatures disturbing effects impeded reliable results (see above). Figure 5.11 depicts the Eyring

5 Immobilization of a Rhodium Catalyst

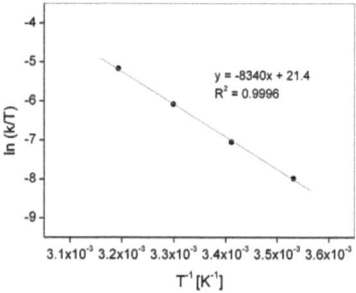

Figure 5.11: Eyring plot for basic conditions.

plot of the investigated temperature range, from which an activation enthalpy $\Delta H^{\ddagger} = 71 \pm 1$ kJ mol^{-1} and an activation entropy $\Delta S^{\ddagger} = -15 \pm 2$ J mol^{-1} K^{-1} were obtained. The values of k and $\Delta S^{\ddagger}$ differ significantly from those found for the homogeneously catalyzed ATH reaction [112]; a discussion follows in Section 5.3.3.

5.3.2 Acidic Reaction Conditions

A comparison of the conversion profiles of the standard experiment and an analogous experiment performed at pH 3.9 (initial value) demonstrated again the significantly higher reaction rate under acidic conditions (Figure 5.12 a). Furthermore, the agreement of computed data and the experimentally determined concentration curve indicated that the first-order dependency of substrate and reaction rate (Equation 2.4.2) was still valid (Figure 5.12 b). Under the acidic conditions, a value of 3.7 ± 0.3 L mol^{-1} s^{-1} for the rate constant k was computed. This is approximately twice as high as the value determined under the standard conditions. The solution pH increased during these experiments from 3.9 to about 4.6.

From an Eyring plot for temperatures between 10 °C and 40 °C at an initial pH of 3.9, a value of 81 ± 1.5 kJ mol^{-1} for $\Delta H^{\ddagger}$ and a value of $+24 \pm 5$ J mol^{-1} K^{-1} for $\Delta S^{\ddagger}$ was determined. The entropy of activation was thus significantly higher than that obtained under the standard conditions; possible mechanistic interpretations are discussed in the following section.

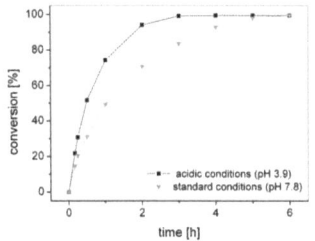
(a) Conversion of acp.
(Standard conditions as in Fig. 5.10).

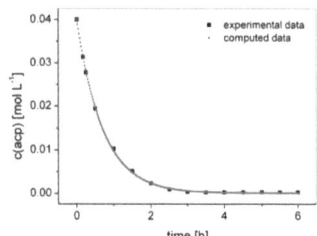
(b) Experimentally determined and computed concentration of acp.

Figure 5.12: ATH of acp under pH-optimized conditions.
Reaction conditions: acp 0.04 M, HCO_2H 0.12 M, HCO_2Na 0.28 M, PB-supported catalyst 1.5 g (~0.105 mM Rh), 40 °C, 1 200 rpm, total reaction volume 100 mL.

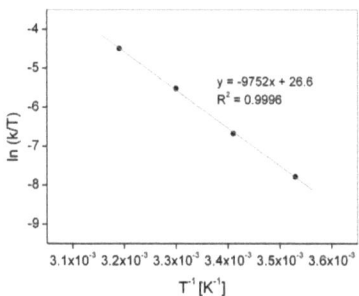

Figure 5.13: Eyring plot for acidic conditions.

5.3.3 Mechanistic Considerations

The second-order kinetics found for the ATH of acetophenone in sodium formate/water solution catalyzed by **22** are in agreement with the findings for the homogeneous untethered Ru-TsDPEN catalyst [112]. The intensive investigations on the ruthenium-catalyzed reaction carried out by Liu and co-workers afforded a well-founded mechanistic concept which explains the variations of the reactivity and enantioselectivity under different reaction conditions (cf. Scheme 2.5 in Section 2.4.4.2). According to this concept the hydrogen transfer proceeds via a highly organized transition state (TS) in which the substrate is coordinated to the metal hydride and the nitrogen proton at the same time (Figure 5.14 a). Under acidic conditions, ring-opening of the chelate ligand due to protonation of the sulfonamide nitrogen is presumed to complicate the formation of the transition state, leading to the observed decrease in reaction rate and enantioselectivity (Figure 5.14 b). These suggestions were supported by a spectroscopic study of the complex under acidic

5 Immobilization of a Rhodium Catalyst

conditions. The low reactivity under basic reaction conditions, however, is explained by an inactive catalyst species formed through the coordination of a hydroxide ion (Figure 5.14 c).

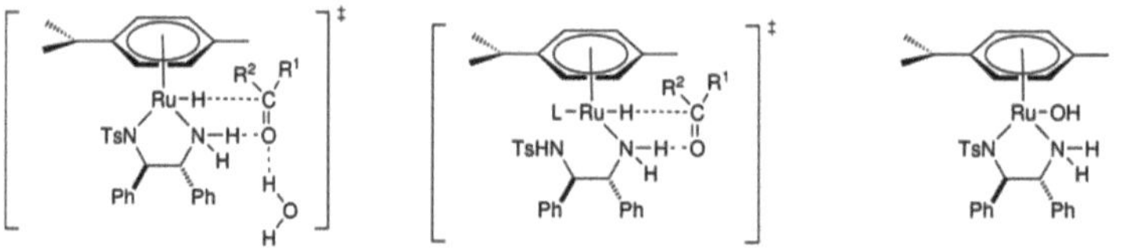

(a) TS under neutral conditions.   (b) TS under acidic conditions.   (c) Inactive species under basic conditions.

Figure 5.14: Ru-TsDPEN species under different conditions.

Because of the concordant rate laws of both catalytic systems under neutral and moderately basic conditions as well as the decline in catalyst activity under strongly basic conditions observed in both cases, an analogous mechanism as proposed for Ru-TsDPEN seems likely for the (immobilized) tethered Rh-TsDPEN catalyst (cf. the middle part of Scheme 5.7 and Figure 2.5). Thus, despite the heterogenization, the different metal and the structural differences, the catalyst is believed to operate in the same way as the homogeneous non-tethered ruthenium analog when those conditions are applied. Nevertheless, in contrast to the Gaussian-like distribution of the TOF around pH 7 which has been found for ruthenium, rhodium, and iridium catalysts based on the TsDPEN ligand [184, 186, 187], the activity of the tethered catalyst applied here peaked at pH 3.9, and an almost constantly high ee was obtained in a pH range between 3.5 and 9.5 (Figure 5.15).

Under the assumption that the structural differences of the ligand do not cause a significant change in the basicity of the nitrogen donor atoms (particularly in the pK_b of the sulfonamide-N), protonation and ring-opening will occur with the tethered rhodium complex in an acidic medium as proposed for the analogous ruthenium complex. However, as long as only one of the nitrogen atoms is protonated — it is most likely that the anionic tosyl-N will be protonated first — the catalyst might be enabled by the tether to keep its steric orientation. An opened ring with a linked cyclopentadienyl moiety might possibly facilitate the formation of the transition state instead of impeding it. These considerations are supported by the higher entropy of activation, which suggests a more flexible transition state under moderately acidic conditions than under basic reaction conditions. Additionally, the tether between the chiral backbone and the cyclopentadienyl moiety provides a plausible explanation for the complete reversibility of the catalyst inhibition under strongly acidic conditions after recovery; even when both nitrogen atoms are protonated and an inactive species is formed, the ligand is connected to the central metal via η^5-coordination (see Scheme 5.7 j). The strength of the η^5-coordination has been demonstrated in a study presented by Perutz et al., where the analogous, but unmodified and non-tethered, Rh-TsDPEN catalyst dissolved in methanol was shown to form triply bridged dimers of rhodium-tetramethylcyclopentadienyl $\langle[\{Cp^*Rh\}_2(\mu\text{-H})(\mu\text{-Cl})(\mu\text{-HCOO})]^+\rangle$ upon the addition of 10 equivalents of formic acid [183]. Thus, the connection to the chiral backbone is released under those conditions whereas the connection to the aryl moiety is not. Consequently, the tethered aryl moiety is assumed to prevent metal leaching and to facilitate reversion to the chelate complex when conditions are

5.3 Kinetic and Mechanistic Investigations

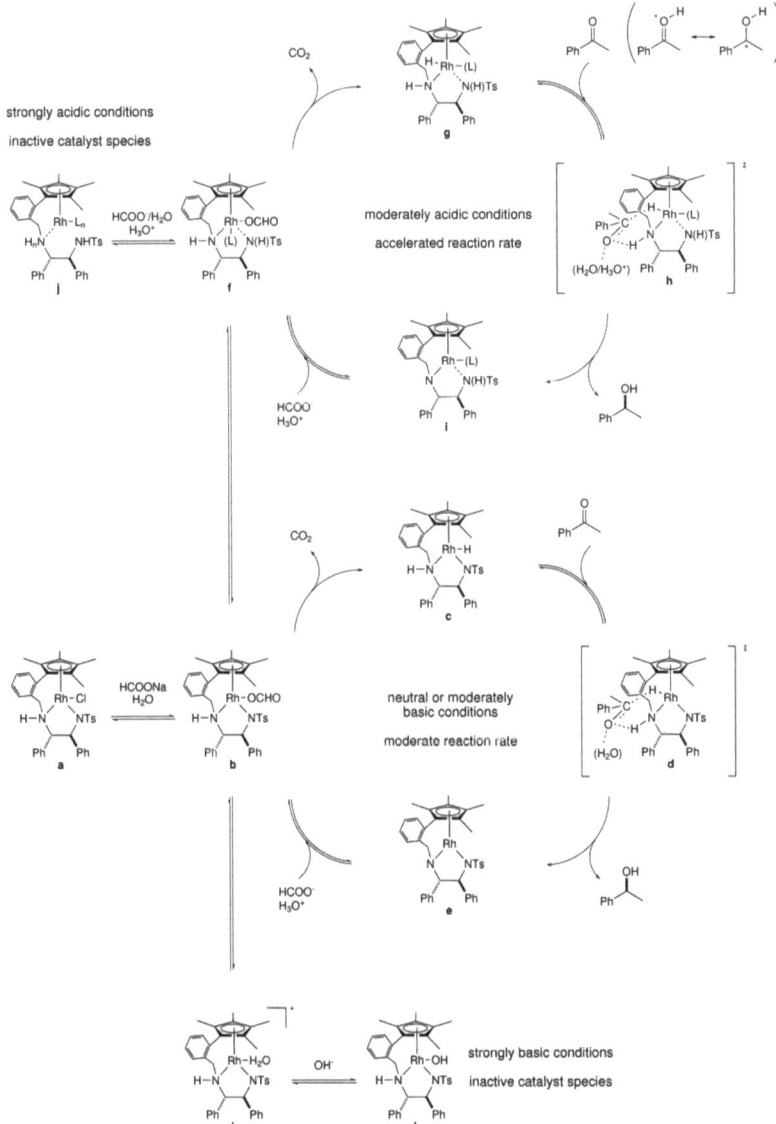

Scheme 5.7: Proposed mechanism under different pH conditions.
Remarks: L might be H_2O, which would lead to a $+1$ charge, or HCO_2^-; dashed bonds are drawn either when undefined interactions between the reactants in the transition state are meant or when there is no certainty about the bond — likewise the atoms/molecules are put in parentheses.

5 Immobilization of a Rhodium Catalyst

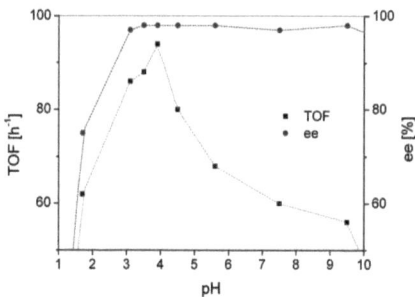

Figure 5.15: Activity and enantioselectivity over pH.

less acidic. Generally, strongly acidic conditions adjusted by the addition of HCl(aq) might effect also inhibition due to chloride coordination to the central metal, but the presence of chloride anions was not mandatory for inhibition of the tethered catalyst, since the activity also decreased without adding HCl(aq) when the ratio of formic acid to sodium formate exceeded 3:7 (cf. Table 5.5, Section 5.2.4).

All attempts to prove the existence of a protonated species via spectroscopic methods, however, were unsuccessful. NMR spectra of dissolved complex **21** in the presence of different equivalents of TFA in CDCl$_3$ were difficult to interpret, but suggested complete decomposition rather than protonation of a nitrogen atom. The difficulties of NMR measurements in this context have been stated in the literature [112]; according to the results reported there, a more sophisticated approach is required to prove the protonation of the catalyst. However, infrared spectra of **21**, which were recorded in the presence of formic acid, also did not provide any reliable evidence of a protonated species.

Nevertheless, some further considerations may be useful in discussing whether ring-opening due to protonation provides a satisfactory explanation for the values of the activation parameters determined from the kinetic experiments. Ring-opening under acidic conditions might provide a transition state which is less ordered than that under basic conditions. For the assessment of the entropy of activation, however, the degrees of freedom of the reactants prior to the formation of the activated complex also have to be discussed. Protonation under acidic conditions which leads immediately to ring-opening would provide a "flexible" catalyst which loses "freedom" upon formation of the transition state. Thus, freedom in absolute terms would not be gained. Nonetheless, protonation of the tosyl-N does not inevitably lead to the release of coordination to the central metal and attended ring-opening [112, 188]. It is possible that the coordinative bond between rhodium and the protonated tosyl-N is not substituted by coordination of a formate anion or a water molecule until the activated complex is formed. In this case, the loss of degrees of freedom due to the interactions between substrate and catalyst in the transition state might be compensated to some extent by the break of the bond between rhodium and tosyl-N. The resulting transition state is depicted in Figure 5.16 a; further options as indicated in Scheme 5.7 (**h**) and Figure 5.16 b/c are discussed below.

Hence, it remains questionable if an increase in entropy, as indicated by the positive sign of $\Delta S^\ddagger$ under

5.3 Kinetic and Mechanistic Investigations

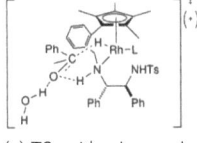
(a) TS with ring-opening and participation of water.

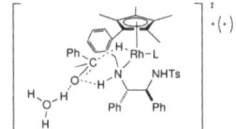
(b) TS with ring-opening and participation of hydronium.

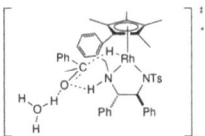
(c) TS without ring-opening with participation of hydronium.

Figure 5.16: Alternative transition states under acidic conditions.

acidic conditions, rather than a (possibly incomplete) compensation of a decrease may be caused through ring-opening. The absolute value and sign of $\Delta S^{\ddagger}$ are calculated from an Eyring plot and depend on the values of the rate constant k at the respective temperatures. The determination of k using immobilized catalyst **22** entails the problem that the exact concentration of substrate in the environment of the catalyst is unknown. An accumulation effect is assumed to occur (see section 5.2.2.1), but no quantitative statement can be given at this time. The calculation of k is therefore based on the total concentration of acp in the reaction mixture, which is quite low. The presumed accumulation of acp in the immediate environment of the catalyst thus leads to a value of k which is probably too high (cf. [189]). The slope in the Eyring plot, which represents the activation enthalpy, is not affected as long as the accumulation of acp is not temperature-dependent. In contrast, the activation entropy is somewhat erroneous, though probably only the absolute values, since accumulation is assumed to occur to the same extent under basic as well as under acidic conditions. Thus, $\Delta\Delta S^{\ddagger} \approx 38$ J mol^{-1} K^{-1} is assumed to be correct. The inhomogeneous concentration of the substrate could explain the great difference to the activation parameters found by Xiao, Liu, and co-workers [112]; using the homogeneous Ru-TsDPEN catalyst for the ATH of acetophenone (under conditions comparable to the "standard reaction conditions"), a $\Delta H^{\ddagger}$ of 53.6 kJ mol^{-1} and a $\Delta S^{\ddagger}$ of -105 J K^{-1} mol^{-1} were found (cf. Section 2.4.4.2). The difference of about -95 J K^{-1} mol^{-1} in the entropy of activation might be best explained by the accumulation effect. In conclusion, a reliable statement that the entropy of activation increases upon the formation of the transition state under acidic conditions cannot be made, but an activated complex, which is less hindered than under basic conditions, is strongly suggested.

Apart from ring-opening as a reason for the higher entropy of activation under acidic conditions, a general explanation for the observed rate acceleration may be acid catalysis. Results of DFT calculations, which were made to elucidate the role of water in the hydrogen transfer mediated by Ru-TsDPEN (in a non-acidic medium), showed that the hydrogen bonding interaction between the amine-H and the carbonyl-O was smaller than without water and that the activation energy was lowered (water made the transition state more "reactant-like" in accordance with the Hammond postulate) [112]. Hydronium ions might intensify this effect, given that the formation of the transition state is not impeded for other reasons. Thus, either protonation promotes the hydrogen transfer or does not interfere with it under moderately acidic conditions (Figure 5.16 a/b), or it does not occur (Figure 5.16 c).

In the first case rate acceleration due to the addition of acid might be in addition to a potential increase

5 Immobilization of a Rhodium Catalyst

in flexibility due to ring-opening (Figure 5.16 b); in the latter it provides the only explanation for a decrease in $\Delta G^{\ddagger}$. The mechanism of the proton transfer might even be completely altered so that the hydronium delivers the proton instead of the catalyst's amine. The amine-H, however, would still be necessary to interact with the carbonyl-O to get the substrate in the right position. It would have to be clarified to which extent the proton transfer may proceed prior to the formation of the transition state. If a transition state as depicted in Figure 5.16 c is true, single protonation would be a sufficient explanation for inhibition, and ring-opening as an explanation for a (relative) increase in $\Delta S^{\ddagger}$ would have to be rejected. Determination of the pK_b values of the unmodified tethered Rh-TsDPEN might provide a first indication.

A further effect that has to be taken into account is solvation, especially when dealing with charged reactants. In this context, it might be useful to discuss the ancillary ligand (L in Scheme 5.7, Figure 5.14 and Figure 5.16) that will saturate the central rhodium ion if ring-opening occurs. The group of Süss-Fink and others have proven the existence of aqua complexes formed from Ru-TsDACH via hydrolysis [185, 190]. Hence, Xiao et al. suggested that water is coordinated to the central metal of Ru-TsDPEN when protonation under acidic conditions leads to an opened ring [79]. However, as they could not isolate an aqua complex, the formate anion was considered as alternative ancillary ligand [112]. Accordingly, L is either H_2O (leading to a +1 charge of the complex) or HCO_2^-. From the above results, however, it seems implausible that L is water when hydronium participates in the transition state at the same time (Figure 5.16 b). This situation would lead from a +1-charged catalyst to a +2-charged activated complex with loss of "freedom" due to electrostriction [191]. Additionally, in case of L being water and the presence of a pre-activated, +1-charged substrate, the frequency of collisions would be reduced according to the kinetic theory of collisions. In summary, a definite statement on the mechanism cannot be made on the basis of the data obtained thus far, and further steps have to be made to provide evidence for one of the mechanistic options considered or even for an entirely new catalytic cycle. These steps include:

- determination of the pK_b of unmodified tethered Rh-TsDPEN catalyst (Figure 5.1) and **21** to identify the pH conditions that lead to protonation and to elucidate differences caused by the structural modification

- determination of the rate of conversion of acp when using the unmodified tethered Rh-TsDPEN catalyst to shed light on the impact of the support

- NMR experiments to elucidate the complex structure in the presence of acids analogous to those described in the literature [112], preferably using the unmodified tethered Rh-TsDPEN catalyst

- DFT calculations to prove the possibility of an enantioselective hydrogen transfer by an "opened" catalyst and to determine the energy barrier for when water is substituted for hydronium.

5.4 Practical Aspects

The experimental results discussed in the preceding sections provide data and information about the specific properties of the system developed here which are required for optimal use on a technical level. However,

5.4 Practical Aspects

from an industrial point of view, several further aspects might be of importance. Studying these aspects does not necessarily provide additional answers to the question of *how* the catalytic system works, but rather to the question of *how well* it works. All results which are considered important in this context are summarized at the end of this chapter.

5.4.1 Application-oriented Experiments

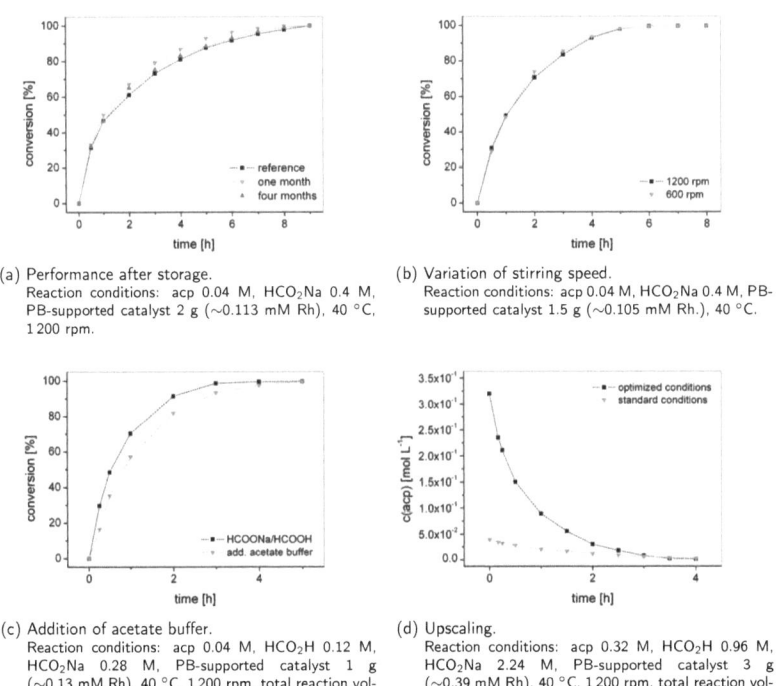

(a) Performance after storage.
Reaction conditions: acp 0.04 M, HCO$_2$Na 0.4 M, PB-supported catalyst 2 g (∼0.113 mM Rh), 40 °C, 1 200 rpm.

(b) Variation of stirring speed.
Reaction conditions: acp 0.04 M, HCO$_2$Na 0.4 M, PB-supported catalyst 1.5 g (∼0.105 mM Rh.), 40 °C.

(c) Addition of acetate buffer.
Reaction conditions: acp 0.04 M, HCO$_2$H 0.12 M, HCO$_2$Na 0.28 M, PB-supported catalyst 1 g (∼0.13 mM Rh), 40 °C, 1 200 rpm, total reaction volume 100 mL (+ HOAc 0.42 M, NaOAc 0.12 M).

(d) Upscaling.
Reaction conditions: acp 0.32 M, HCO$_2$H 0.96 M, HCO$_2$Na 2.24 M, PB-supported catalyst 3 g (∼0.39 mM Rh), 40 °C, 1 200 rpm, total reaction volume 100 mL; standard conditions as in Fig. 5.10.

Figure 5.17: Technically relevant tests.

The high stability of **22** and analogous catalysts in reactions performed in water and air was demonstrated in this and other studies [176]. It was found that a period of about four months of storage in air in the refrigerator did not affect the catalyst performance (Figure 5.17 a). It was thus concluded that durability does not pose a problem with the supported tethered Rh-TsDPEN catalyst.

On a technical scale, stirring contributes considerably to the cost of a given process since it influences the dimensioning of the reactor as well as the energy consumption. Reducing the stirring speed may thus be a useful approach to reduce costs. As discussed previously, the aggregation of beads leads to a decrease

5 Immobilization of a Rhodium Catalyst

in the rate and reproducibility of the reaction. A high stirring speed of 1 200 rpm was therefore applied in all experiments. Under certain conditions, however, the stirring speed can be significantly reduced without affecting the performance of the reaction. Figure 5.17 b shows that there is no significant difference in the conversion of acp over time in the standard experiment performed at 1 200 rpm and at 600 rpm. Whether the suspension is homogeneous or not, though, also depends on the number of beads and the concentration of the substrate. Hence, the optimal stirring speed has to be determined individually for each approach.

The pH was found to be a decisive parameter for the performance of the catalyst, and the activity (TOF) peaked at an initial pH of about 4. Since the basicity of the reaction medium increased with increasing conversion, the effect of an additional buffer with a higher capacity than HCO_2H/HCO_2Na was tested. Through the addition of small amounts of acetate buffer, the increase of the pH was reduced; when an initial pH of 3.9 was set, a value of 4.2 was determined at the end of the reaction instead of 4.6 (conversion >99 %). The reaction rate, however, was reduced rather than improved (Figure 5.17 c). This might be ascribed to a disfavorable interaction between acetate and catalyst. Consequently, a more suitable buffer system might be capable of improving the reaction, but it also might be challenging to find the optimal system at the optimal concentration. Since the expected improvement was marginal, no further tests on this matter were carried out.

However, the efficiency of the reaction can be improved and transferred to a more technically relevant level if the concentrations of catalyst and substrate as well as the S/C are increased. The use of additional beads in the solution was limited, since a homogeneous dispersion was required to ensure reproducible conditions and a fast conversion. At the same time, aggregation occurred mainly at a higher substrate concentration when a smaller number of beads was employed. A compromise was reached using beads with higher loading (13 $\mu mol\, g^{-1}$ instead of 7 $\mu mol\, g^{-1}$) while increasing the amount by a factor of two (from 1.5 g to 3 g), whereas the concentration of acetophenone was increased by a factor of eight to 0.32 M (Figure 5.17 d). Under these conditions, little aggregation occurred at a stirring speed of 800 rpm, and an STY of $10 \pm 1\, g\, L^{-1}\, h^{-1}$ with an average TOF of $200 \pm 10\, h^{-1}$ was achieved.

5.4.2 Discussion on the Efficiency

It was demonstrated that the immobilized Rh(III)-catalyst (**22**) developed in this study can be applied in ATH reactions of various aryl ketones (cf. Section 5.2.2.3); all performance tests, however, were conducted exclusively using acetophenone as the substrate. Thus, the following considerations on the strengths and weaknesses of the catalytic system and its potential for an industrial application are based on this benchmark compound, taking into account the industrial requirements discussed in Section 2.2 and Section 2.4.6.

Generally, the selectivity of ATH reactions is 100 %, because there is no generation of side products from the prochiral substrate. Since in sodium formate/formic acid/water complete conversion can be achieved, the theoretical yield is also 100 %. The enantioselectivity achieved with **22** is about 98 % ee of 1-(*S*)-phenylethanol under optimal reaction conditions (40 °C, pH 3.9). This is in the range of the best results found in the literature and significantly exceeds the minimum requirement of 90 % ee in pharmaceutical production processes. Furthermore, a TON of at least 1 000 is required for the small scale fabrication of high added value products. This requirement is met by catalyst reuse which provided a TTN of over 1 200

5.4 Practical Aspects

in eight consecutive runs with a total loss of metal of about 5 %. Upscaling experiments, however, revealed that even a single run performed at a higher S/C can provide a TON of more than 800. For the TOF, which should be higher than 500 h^{-1} at 95 % conversion, a value of 200 h^{-1} (at >99 % conversion) was achieved, hence there is a need for improvement. Both TOF and TON may be enhanced by the application of a higher substrate concentration; the TOF will also increase when a higher catalyst concentration is applied in addition to a higher substrate concentration. The problem of bead aggregation will therefore have to be addressed; the addition of a polar organic co-solvent such as methanol may be a starting point. An increased substrate concentration will also improve the STY, which was about 10 g L^{-1} h^{-1}.

A drawback of the functionalized support materials is the the low surface area, which limits the loading with amino groups (15–30 µmol g^{-1}) and consequently the catalyst loading. A molecular weight of the heterogeneous catalyst of about 10 kDa per mole of active site cannot be achieved when these materials are applied. Particles that are porous and smaller in diameter are required if this is truly a decisive criterion. One advantage of the catalytic system present here (**22**), however, is its robustness. The stability to air as well as the high mechanical stability allow for the use of simple reaction and filtration setups, and the insensitivity to carbon dioxide makes gas disengagement unnecessary.

Nevertheless, the question of efficiency is not settled. Catalyst immobilization — in particular heterogenization — is generally associated with increased effort; a statement on the advantage of an immobilized catalyst over its free analog should therefore consider the potentially higher preparation costs. The established key figures which are used to compare catalyst and process performances, however, do not include such factors. This might be for various reasons: costs (as a dimension of "effort") are usually difficult to determine, most often do not play a role in scientific publications, are kept secret in industry, etc. If however they are known, they should be considered by using a modified productivity indicator, which could be referred to as "catalyst cost efficiency" (CCE, Equation 5.4.1), borrowing a term coined by Campbell [192].

Since the catalyst developed in this study will be commercially available from *Strem Chemicals Inc.*, this key figure may be used to decide whether an application could be advantageous or not. Of course, the overall process is not taken into account, and further considerations, e. g., on the effort and costs required for separation, will have to be made. If data on the catalyst costs are not available, the "catalyst efficiency" (CE) could be estimated using Equation 5.4.2.

$$\mathbf{CCE} \; [\text{mol} \, \text{\euro}^{-1}] = \frac{\text{converted starting material [mol]}}{\text{amount of catalyst employed [mol]} \cdot \text{costs of catalyst [\euro mol}^{-1}]} \tag{5.4.1}$$

$$\mathbf{CE} = \frac{\text{converted starting material [mol]}}{\text{amount of catalyst employed [mol]}} \cdot \frac{\text{yield}}{\text{number of synthetic steps}} \tag{5.4.2}$$

This latter approach may be oversimplified as it does not differentiate among the synthetic steps, but it does consider the resources employed (number of synthetic steps) and the yield obtained in the preparation of the catalyst. For example, three linear steps are required to prepare the unmodified tethered Rh-TsDACH complex (Figure 5.1) when chiral DACH is considered the starting compound (not taking into account the preparation of the tether). Presuming a yield of 90 % in the first step (preparation of TsDACH), the overall yield should be about 15 % [95, 177]. Analogous calculation of the overall yield in the preparation of **22** gives about 11% in nine steps, not taking into account the synthesis of the tether and the preparation of the support. When an S/C of 200 is applied and complete conversion in a single run is achieved with the homogeneous Rh-TsDACH catalyst, and a TTN of 1 250 is achieved with the supported catalyst under otherwise comparable conditions, CE values of 10 for Rh-TsDACH and 15 for **22** are obtained. Although not all preparation "costs" and application advantages are taken into account, this might at least demonstrate that heterogenization is becoming a serious alternative to homogeneous catalytic systems.

Apart from economic aspects, catalytic transfer hydrogenation reactions offer a number of benefits over alternative methods according to the principles of green chemistry and green engineering, e. g., the use of safe solvents and auxiliaries as well as safe processes. Nevertheless, ATH reactions are usually not considered atom economical [28]. This is mainly because the hydrogen source itself is transformed into "waste" during the process. The determination of the E-factor or another key figure which assesses the atom economy by taking into account the waste generated is challenging (see Section 2.2.3). An estimation of the E-factor following the simplified approach, i.e., by only taking into account the reagents and the desired product (not taking into account the catalyst, solvent, energy etc.) gives a value of about 0.7 when the reaction is performed with a stoichiometric amount of sodium formate, but 5.7 when 10 equivalents are applied and non-transformed sodium formate is considered as waste. analogous data of alternative methods are required to compare these values; a thorough determination of the ecological impact, however, should consider the overall process.

6 General Conclusion and Outlook

Asymmetric transfer hydrogenation is becoming a more commonly applied industrial method. The intensive research performed during the last 20–30 years has afforded substantial progress, including the development of highly efficient catalysts applicable to different reaction media and different hydrogen donors. Not only are the operational aspects of this method persuasive (high safety and simplicity), but the overall performance has become competitive compared to other approaches. Additionally, many attempts have been made to immobilize ATH catalysts to address the problem of efficient catalyst separation and reuse, and promising results have been obtained. Only very few studies in this field, however, have focused on application-oriented data and kinetic parameters.

The aim of this work was the preparation and study of immobilized catalysts for the ATH of prochiral ketones. Heterogenization of outstanding homogeneous catalysts was chosen as the immobilization strategy, and kinetic studies were performed with the goal of obtaining a fundamental understanding of the operating mode of the catalytic systems. Two transition metal complexes were subjected to synthetic modification in order to introduce linkers suitable for the covalent attachment to solid polymer supports. These supports were prepared via Molecular Surface Engineering, i. e., a "functional matrix" containing (protected) primary amines was grafted from either polypropylene membranes or polyethylene sinter chips and beads.

In a first attempt, a diphosphine/diamine (PNNP) ligand was modified via a new nine-step convergent synthesis process to introduce a carboxylic linker [86]. For the purpose of ATH operations, the ligand could either be used with ruthenium(II) in isopropyl alcohol solutions or with iridium(III) — after in situ generation — in aqueous media [168]. However, all attempts at covalent attachment of both the ligand and the corresponding ruthenium complex failed. Another attempt was made with the linkage of a modified rhodium(III) complex based on a variant of the monotosylated diphenylethylenediamine (TsDPEN) ligand, which features a linked η^5-cyclopentadienyl unit [95]. A carboxylic linker was introduced in a joint project with the Haag Research Group, and after metal insertion and successful attachment to the supports, a new immobilized catalyst, the supported tethered Rh-TsDPEN (**22**), was applied in the ATH of aryl ketones. Comparison of the catalytic performance in the sodium formate/water system in air showed that the enantioselectivity in the transfer hydrogenation of acetophenone and related substrates was as high as or even better than reported for the unmodified analogous catalysts in the literature [79, 177]. In terms of reaction rate, the polymer-supported catalyst (**22**) was found to be superior even in comparison to the monomeric analog (**23**) in the concentration range investigated. This is ascribed to the properties of the support material; an accumulation of the substrate in the microenvironment of the catalyst is likely to occur, leading to an enhanced rate at low substrate concentrations. A successful reuse after simple filtration was achieved when an acidic mixture of sodium formate and formic acid in water was used.

6 General Conclusion and Outlook

Kinetic investigation revealed that the conversion of acetophenone to phenylethanol follows a first-order dependency of the reaction rate on both the substrate and the catalyst concentration. It was therefore suggested that the operating mode of the immobilized tethered catalyst is similar to that proposed for untethered metal-arene-TsDPEN catalysts in the moderately basic reaction medium sodium formate/water [112]. However, considerable differences in the catalyst performance compared to previously reported results were noted when the reaction conditions were changed: increased activity and reusability without loss of enantioselectivity of the immobilized system were observed in the acidic $HCO_2H/HCO_2Na/H_2O$ medium, but the second-order rate law was found to still be valid. Thus, the optimal pH for this catalyst is considerably below that reported for the untethered analogs (pH ~4 and pH ~7, respectively) [184, 186, 187]. Additionally, emerging carbon dioxide was found to accelerate the reaction rather than inhibiting it, due to its buffer capacity. These findings are presumably ascribable to the structural superiority of the immobilized complex, effected by the tether between chiral backbone and cyclopentadienyl ligand. Comparison of the activation parameters of reactions carried out in moderately basic and moderately acidic media revealed a significant difference in the entropy of activation ($\Delta S^{\ddagger}_{acidic} - \Delta S^{\ddagger}_{basic} \approx 38$ J mol^{-1} K^{-1}). The underlying effect of the rate enhancement under acidic conditions is thus presumed to be an altered mechanism in which — apart from possible solvation effects — either ring-opening or acid catalysis or both provide a "less ordered" transition state. If ring-opening occurs, the tether might provide an explanation why this catalyst, in contrast to untethered variants, is capable of keeping its steric orientation. Further investigations including the spectroscopic analysis of the homogeneous analog as well as DFT calculations are required to elucidate these considerations. A deeper understanding of the mechanism may help in developing even more efficient catalysts in the future.

The accumulation of the substrate is presumed to derive from a better solubility of the substrate in the outer layer of the support than in the solvent. This is only the case when a polar solvent is used. The significantly reduced rate observed when the reaction was run in organic media instead of water might be a further indication, although a possible involvement of water in the reaction mechanism also has to be taken into account. However, substrate accumulation would be a viable way of enhancing the rate of ATH reactions carried out in IPA, since these reactions are mostly performed at high dilution to shift the equilibrium. The development of the PNNP-based catalyst was indeed focused on the 2-propanol system, but in light of the above findings, the effectiveness of this approach is called into question. As long as the support does not attract the substrate to at least the same extent as the solvent (and to a higher extent than the product), one of the main drawbacks usually associated with heterogenization — a strong decrease in catalyst activity — would be inevitable. Since even homogeneously applied Ru-PNNP shows a relatively low activity, the effort of modification and immobilization would not be worth the cost, neither economically nor ecologically. In principle, the polarity and hydrophilicity/hydrophobicity balance of the surface of the supports can be adjusted via Molecular Surface Engineering, but the range of adjustments is, of course, limited. Thus, the immobilization strategy applied here seems to be inefficient for catalysts that can only be used in medium polar or nonpolar media.

Nonetheless, the application of the immobilized Rh(III)-catalyst (**22**) to ATH reactions of aryl ketones in aqueous media provided promising results, though there is certainly plenty of room for optimization. As this is the first report of catalyst immobilization together with kinetic investigation applying a tethered version of the TsDPEN ligand as well as the first time that an immobilized ATH catalyst will be commercially available, it will hopefully be a starting point for further studies. Future work will include both the optimization and deeper understanding of the system developed as well as research into further aspects, particularly with respect to environmental issues of an industrial application. The use of recyclable catalysts already complies with several principles of green chemistry and green engineering. To further improve the ecological impact, the problems of a cycle of matter and a commercial "afterlife" will possibly be addressed in a joint project with an industrial partner. The commercial application of a continuously operated reaction system is currently being tested. Furthermore, an ATH process in combination with a membrane-based separation of the organic products from the aqueous reaction medium is envisaged. Since analogous ruthenium complexes have been found to catalyze different reactions in addition to ATH, e.g., the asymmetric Michael-Addition [193, 194], the range of applications could be further tested. A combined chemo-/biocatalytical application is already under investigation at Delft University of Technology

7 Experimental

Unless preparative details are provided, all reagents were commercially available from standard suppliers and used without further purification. Compounds known in the literature are referenced where appropriate in the procedure. Reactions which required anhydrous and inert conditions were conducted in oven-dried apparatus and in an atmosphere of argon (5.0) using standard Schlenk techniques and dry solvents; addition of matter or sampling was routinely performed in a counter current flow of argon in such reactions. Column chromatography was performed with silica gel (60 mesh, 0.06–0.2 mm particle size). Synthetic workup and washing procedures were performed with deionized water, catalytic testing was performed using water purified by reverse osmosis. Thin layer chromatography (TLC) was used to monitor reactions where appropriate. Visualization of the plates was by 254 nm UV light and/or iodine staining.

Nuclear magnetic resonance (NMR) spectra were recorded at room temperature (rt), and chemical shifts (δ), measured in parts per million (ppm), are reported relative to the tetramethylsilane signal ($\delta = 0.00$ ppm, external standard for ^{1}H- and ^{13}C-NMR measurements) and the phosphorous acid signal ($\delta = 0.00$ ppm, external standard for ^{31}P-NMR measurements), respectively. Multiplicities are denoted as singlet (s), doublet (d), doublet of doublets (dd), triplet (t), pentet (p), and multiplet (m), and where required prefixed br (broad). Infrared (IR) spectra were recorded either from pure solids or from thin films of solutions in chloroform using attenuated total reflection (ATR) technique. IR absorptions are reported in wavenumbers (ν), measured in cm^{-1}. High resolution mass spectrometry (HRMS) was performed by electrospray ionization (ESI), either via direct injection or coupled with high performance liquid chromatography (HPLC). The calculation of the "mass to elementary charges ratio" (m/z) as well as the interpretation of the high resolution mass spectra were performed with *Xcalibur Qual Browser*, version 2.0.7.

The determination of the conversion in catalysis testing was performed via gas chromatography (GC) using a flame ionization detector (FID) or a coupled mass spectrometer (MS) equipped with an electron ionization (EI) device. The GC-MS system was calibrated for the determination of the conversion of acetophenone to 1-phenylethanol, which was calculated by dividing the peak area of the ion fragments produced from acp by the sum of the ion fragments of both compounds. Computational data were obtained by the use of *Madonna* software, version 8.3.14.

7.1 Equipment

7.1.1 Instrumentation

- GC (analysis of the headspace gas in catalysis experiments): *Shimadzu GC-2014* gas chromatograph equipped with *ValcoPLOT* molsieve, *HayeSep D* and *HayeSep Q* columns, and an FID as well as a thermal conductivity detector.

- GC (analysis in catalysis experiments): *Varian CP-3800* equipped with an FID detector, a *Varian CP-8400 AutoSampler*, test column (*CP-Sil B CB*; 15 m x 0.25 mm x 0.25 μm), or chiral column (*CP-Chirasil-Dex CB*; 25 m x 0.25 mm x 0.25 μm).

- GC-MS (analysis in catalysis experiments): *Hewlett Packard HP 6890 Series* gas chromatography system equipped with a *HP 5975 Mass Selective Detector* using electron ionization (EI) at 70 eV and a *Gerstel MPS 2L* auto sampler. A *Machery-Nagel Optima Wax* column (30 m x 0.32 mm x 0.5 μm) was used for determination of the conversion, and for determination of the enantiomeric excess a *Supelco Beta DEX 110* column (30 m x 0.25 mm x 0.25 μm) was used.

- GC-MS (analysis in synthesis): *Hewlett Packard HP 6890 Series/HP 5975 Mass Selective Detector* as specified above, equipped with an *HP-5MS* GC column (30 m x 0.25 mm x 0.25 μm).

- HRMS: *Thermo Scientific Orbitrap LTQ XL* (spray voltage 5 kV, source temperature 275 °C, solvent methanol + 0.1 % formic acid, flow rate 200 μL min^{-1}); HPLC conditions: *Agilent Eclipse XDB-C18* column (4.6 x 150 mm, 5 μm), eluent 1 (H_2O + 0.025 % HCO_2H), eluent 2 (MeOH + 0.025 % HCO_2H), flow 1.0 mL min^{-1}.

- ICP-OES: *Varian 715-ES*.

- ICP-MS: *Thermo Fisher Element 2*.

- FT-IR: *Perkin Elmer Spectrum one* with *Universal ATR Sampling Accessory*.

- Microwave Digester: *CEM Discover SP-D*.

- NMR (^{1}H, ^{13}C): *Bruker Avance 400*.

- NMR (^{31}P): *Bruker AC-F 200*.

- UV-Vis: *Hitachi U-3410 Spectrophotometer*.

7.1.2 Laboratory Equipment

- Mechanical stirrer: *Heidolph RZR 2051 control*.

- Thermostat: *LAUDA Ecoline Staredition RE 312*.

- Vacuum pump: *ILMVAC diaphragm pump*.

- Magnetic stirrer: *HEIDOLPH MR Hei-Standard* magnetic stirrer with heating equipped with temperature control *EKT Hei-Con*.

- Temperature sensor: *testo 720*.

7 Experimental

- pH meter: *HANNA Instruments pH 211*.
- Reaction vessel: *Rettberg* jacketed 100-mL flask.
- Rotary evaporator: *Heidolph Laborota 4000* equipped with *Vacuubrand MZ 2C* diaphragm vacuum pump.
- Reverse osmosis water purification system: *Millipore Elix 5*.
- Ion exchange system: *SD 2800*.

7.2 Synthesis Procedures

7.2.1 Synthesis of Ru-PNNP

7.2.1.1 Hex-5-ynoic acid *tert*-butyl ester (2)

Different from the procedure used by Bartoli et al. [195], the introduction of the *tert*-butyl protecting group into the linker moiety was performed via Steglich esterification [196]. 5-Hexynoic acid (10 mmol, 1.12 g), *tert*-butanol (30 mmol, 2.22 g), and 4-(dimethylamino)pyridine (DMAP; 0.6 mmol, 0.073 g) were dissolved in 10 mL of dichloromethane (DCM). N,N'-Dicyclohexylcarbodiimide (DCC; 11 mmol, 2.27 g) was carefully added while the flask was cooled in ice water. After 4 h of stirring at room temperature, the reaction was completed (GC-MS control), and precipitated dicyclohexylurea (DCU) was filtered off. The filtrate was concentrated under reduced pressure and purified via column chromatography (eluent: DCM) to yield a colorless oil ($Y_2 = 70\%$).

^{1}H-NMR (400 MHz, CDCl$_3$): δ (ppm) = 2.31 (t, 2 H, C$\underline{H}_2$COOtBu, J = 7.4 Hz), 2.21 (dt, 2 H, C$\underline{H}_2$C$_{alkynyl}$, J = 2.6 Hz, J = 7.0 Hz), 1.93 (t, 1 H, C$_{alkynyl}$$\underline{H}$, J = 2.7 Hz), 1.77 (p, 2 H, CH$_2$C$\underline{H}_2$CH$_2$, J = 7.2 Hz), 1.41 (s, 9 H, C$\underline{H}_3$).

^{13}C-NMR (100 MHz, CDCl$_3$): δ (ppm) = 172.3 ($\underline{C}$OOtBu), 83.3 ($\underline{C}_{alkynyl}$CH$_2$), 80.1 [$\underline{C}$(CH$_3$)$_3$], 68.8 ($\underline{C}_{alkynyl}$H), 34.1 ($\underline{C}$H$_2$COOtBu), 28.0 (3 C, $\underline{C}$H$_3$), 23.7 (CH$_2$$\underline{C}H_2CH_2$), 17.7 ($\underline{C}H_2C_{alkynyl}$).

HRMS (ESI): m/z calculated for C$_{10}$H$_{17}$O$_2$ ([M+H]$^+$) = 169.12231 (100 %), 170.12566 (10.82 %); m/z found = 169.12181 (100 %), 170.12517 (9.10 %).

GC-MS (EI): m/z = 153.0 ([M − CH$_3$]$^+$), 112.0 ([M − C$_4$H$_8$]$^+$), 95.0 ([M − C$_4$H$_9$O]$^+$), 67.0 ([M − C$_5$H$_9$O$_2$]$^+$), 57.1 ([C$_4$H$_9$]$^+$).

7.2.1.2 2-Bromo-5-iodophenylmethanol (4)

4

Borane dimethyl sulfide complex (24 mmol, 1.94 g) was added over 5 min to a solution of 2-bromo-5-iodobenzoic acid (10 mmol, 3.27 g) in THF (40 mL). After stirring for 4 h at room temperature, no further conversion was observed (TLC control). Water (25 mL) was added in order to hydrolyze remaining borane components, and the mixture was stirred for 15 min. After extraction with DCM, the organic phase was dried over MgSO$_4$, and the solvent was removed under reduced pressure. The product, a white solid, was further purified by recrystallization from hexane; the overall yield was 80 %.

^{1}H-NMR (400 MHz, MeOH-d$_4$): δ (ppm) = 7.85 (d, 1 H, C$_\phi$H, J = 2.1 Hz), 7.47 (dd, 1 H, C$_\phi$H, J = 2.2 Hz, J = 8.3 Hz), 7.26 (d, 1 H, C$_\phi$H, J = 8.3 Hz), 4.58 (s, 2 H, CH$_2$OH).

^{13}C-NMR (100 MHz, MeOH-d$_4$): δ (ppm) = 142.7 (C$_\phi$CH$_2$OH), 137.2–133.6 (3 C, C$_\phi$H), 120.9 (C$_\phi$Br), 91.9 (C$_\phi$I), 62.5 (CH$_2$OH).

GC-MS (EI): m/z = 313.9/311.9 ([M]$^+$), 232.9 ([M − Br]$^+$).

7.2.1.3 2-Bromo-5-(hex-5-ynoic acid *tert*-butyl ester)-phenylmethanol (5)

5

Under inert and anhydrous conditions, compound **4** (10 mmol, 3.13 g) and compound **2** (12 mmol, 2.02 g) were dissolved in THF (30 mL). While the solution was beeing stirred at room temperature, PdCl$_2$(PPh$_3$)$_2$ (0.3 mmol, 0.211 g), CuI (0.5 mmol, 0.095 g), and triethylamine (25 mmol, 3.5 mL) were added in succession. After 5 h no further conversion was observed (TLC control). Precipitated triethylamine hydroiodide was filtered off, and the filtrate was concentrated by solvent evaporation under reduced pressure. The residue was dissolved in DCM and washed with a saturated aqueous solution of NH$_4$Cl. The organic phase was then dried over MgSO$_4$, and solvents were removed under reduced pressure. Column chromatography (eluent: DCM/hexane, 1/1) was performed for purification yielding the product in 62 %

yield as a yellow-orange oil of moderate viscosity.

^{1}H-NMR (400 MHz, CDCl$_3$): δ (ppm) = 7.51 (d, 1 H, C$_\phi$H, J = 2.1 Hz), 7.43 (d, 1 H, C$_\phi$H, J = 8.2 Hz), 7.16 (dd, 1 H, C$_\phi$H, J = 2.1 Hz, J = 8.2 Hz), 4.69 (d, 2 H, CH$_2$OH, J = 5.6 Hz), 2.44 (t, 2 H, CH$_2$COOtBu, J = 7.0 Hz), 2.38 (t, 2 H, CH$_2$C$_{alkynyl}$, J = 7.5 Hz), 2.23 (t, 1 H, OH, J = 6.0 Hz), 1.87 (p, 2 H, CH$_2$CH$_2$CH$_2$, J = 7.2 Hz), 1.45 (s, 9 H, CH$_3$).

^{13}C-NMR (100 MHz, CDCl$_3$): δ (ppm) = 172.6 (COOtBu), 139.8 (C$_\phi$CH$_2$OH), 132.4–131.7 (3 C, C$_\phi$H), 123.5 (C$_\phi$Br), 121.6 (C$_\phi$C$_{alkynyl}$), 90.4 (C$_{alkynyl}$CH$_2$), 80.4 (C$_{alkynyl}$C$_\phi$), 80.3 [C(CH$_3$)$_3$], 64.7 (CH$_2$OH), 34.5 (CH$_2$COOtBu), 28.1 (3 C, CH$_3$), 24.0 (CH$_2$CH$_2$CH$_2$), 18.9 (CH$_2$C$_{alkynyl}$).

HRMS (ESI): m/z calculated for C$_{17}$H$_{22}$BrO$_3$ ([M+H]$^+$) = 353.07468 (100%), 354.07804 (18.4%), 355.07264 (97.3%), 356.07599 (17.9%); m/z found = 353.07416 (100%), 354.07739 (13.8%), 355.07211 (95.8%), 356.07538 (13.4%).

7.2.1.4 2-Bromo-5-(hex-5-ynoic acid *tert*-butyl ester)-benzaldehyde (6)

6

Equivalent amounts (10 mmol) of compound **5** (3.53 g) and Dess–Martin periodinane (4.24 g) [197] were dissolved in DCM (30 mL). Stirring at room temperature for 10 h gave a brownish mixture which turned transparent by addition of 1.3 M sodium hydroxide solution (40 mL). After extraction with diethyl ether, the organic phase was dried over MgSO$_4$ and concentrated under reduced pressure. Further purification via column chromatography (eluent: DCM) resulted in a clear yellow oil in 68% yield.

^{1}H-NMR (400 MHz, CDCl$_3$): δ (ppm) = 10.30 (s, 1 H, CHO), 7.89 (d, 1 H, C$_\phi$H, J = 2.2 Hz), 7.56 (d, 1 H, C$_\phi$H, J = 8.3 Hz), 7.42 (dd, 1 H, C$_\phi$H, J = 2.2 Hz, J = 8.3 Hz), 2.46 (t, 2 H, CH$_2$COOtBu, J = 7.0 Hz), 2.39 (t, 2 H, J = 7.4 Hz, CH$_2$C$_{alkynyl}$), 1.88 (p, 2 H, CH$_2$CH$_2$CH$_2$, J = 7.2 Hz), 1.45 (s, 9 H, CH$_3$).

^{13}C-NMR (100 MHz, CDCl$_3$): δ (ppm) = 191.2 (CHO), 172.4 (COOtBu), 137.7 (C$_\phi$H), 133.8 (C$_\phi$H), 133.3 (C$_\phi$CHO), 132.8 (C$_\phi$H), 125.7 (C$_\phi$Br), 124.2 (C$_\phi$C$_{alkynyl}$), 92.0 (C$_{alkynyl}$CH$_2$), 80.5 (C$_{alkynyl}$C$_\phi$), 79.2 [C(CH$_3$)$_3$], 34.4 (CH$_2$COOtBu), 28.1 (3 C, CH$_3$), 23.9 (CH$_2$CH$_2$CH$_2$), 18.8 (CH$_2$C$_{alkynyl}$).

HRMS (ESI): m/z calculated for C$_{17}$H$_{20}$BrO$_3$ ([M+H]$^+$) = 351.05903 (100%), 352.06239 (18.4%), 353.05699 (97.3%), 354.06034 (17.9%); m/z found = 351.05844 (100%), 352.06168 (15.6%), 353.05634 (95.8%), 354.05963 (13.4%).

7.2 Synthesis Procedures

7.2.1.5 5-(Hex-5-ynoic acid *tert*-butyl ester)-2-(diphenylphosphino)benzaldehyde (7)

7

Under inert and anhydrous conditions, compound **6** (10 mmol, 3.51 g) was dissolved in toluene (20 mL). Pd(PPh$_3$)$_4$ (0.07 mmol, 0.081 g), dissolved in 10 mL of toluene, diphenylphosphine (13 mmol, 2.42 g), and triethylamine (15 mmol, 2.1 mL) were added successively. The reaction mixture was refluxed for 3 h, then precipitated triethylamine hydrobromide was filtered off. The filtrate was washed with an aqueous solution of NH$_4$Cl and brine. After separation from the aqueous phase, the organic phase was dried over MgSO$_4$ and concentrated under reduced pressure. The product was obtained in 56 % yield after column chromatography (eluent: DCM) as an viscous, bright yellow oil.

^{1}H-NMR (400 MHz, CDCl$_3$): δ (ppm) = 10.43 (d, 1 H, C$\underline{H}$O, J = 5.3 Hz), 7.97 (m, 1 H, C$_\phi\underline{H}$), 7.45–7.24 (m, 11 H, C$_\phi\underline{H}$), 6.89 (m, 1 H, C$_\phi\underline{H}$), 2.48 (t, 2 H, C$\underline{H_2}$COOtBu, J = 7.0 Hz), 2.40 (t, 2 H, C$\underline{H_2}$C$_{alkynyl}$, J = 7.4 Hz), 1.89 (p, 2 H, CH$_2$C$\underline{H_2}$CH$_2$), 1.45 (s, 9 H, C$\underline{H_3}$).

^{13}C-NMR (100 MHz, CDCl$_3$): δ (ppm) = 191.10 (d, 1 C, $\underline{C}$HO, ^{3}J = 18.4 Hz), 172.4 ($\underline{C}$OOtBu), 140.4 (d, 1 C, $\underline{C}_\phi$CHO, ^{2}J = 26.7 Hz), 138.3 (d, 1 C, $\underline{C}_\phi$P, J = 14.6 Hz), 136.1 ($\underline{C}_\phi$H), 135.9 (d, 2 C, $\underline{C}_\phi$P, J = 9.5 Hz), 134.1–133.8 (5 C, $\underline{C}_\phi$H), 133.7 (d, 1 C, $\underline{C}_\phi$H, J = 3.7 Hz), 129.2–128.8 (6 C, $\underline{C}_\phi$H), 124.9 ($\underline{C}_\phi$C$_{alkynyl}$), 92.0 ($\underline{C}_{alkynyl}$CH$_2$), 80.4 ($\underline{C}_{alkynyl}$C$_\phi$), 79.9 ($\underline{C}$(CH$_3$)$_3$O), 34.4 ($\underline{C}$H$_2$), 28.1(3 C, $\underline{C}$H$_3$), 24.0 ($\underline{C}$H$_2$), 18.9 ($\underline{C}$H$_2$).

^{31}P-NMR (80 MHz, CDCl$_3$): δ (ppm) = -13.13.

HRMS (ESI): m/z calculated for C$_{29}$H$_{30}$O$_3$P ([M+H]$^+$) = 457.19271 (100 %), 458,19606 (31.4 %), 459,19942 (4.8 %); m/z found = 457.19468 (100 %), 458.19800 (30.2 %), 459.20123 (4.1 %).

7.2.1.6 2-(Diphenylphosphino)benzaldehyde (9)

2-(Diphenylphosphino)benzaldehyde **9** is commercially available from different suppliers. Nevertheless, it was synthesized in an analogous manner as **7** under inert and anhydrous conditions. 2-Bromobenzaldehyde (10 mmol, 1.85 g) was dissolved in 20 mL of toluen. Pd(PPh$_3$)$_4$ (0.07 mmol, 0.081 g), dissolved in 10 mL of toluene, diphenylphosphine (13 mmol, 2.42 g), and triethylamine (15 mmol, 2.1 mL) were added successively. The reaction mixture was refluxed for 3 h, then precipitated triethylamine hydrobromide was filtered off. The filtrate was washed with an aqueous solution of NH$_4$Cl and brine. After separation from

7 Experimental

9

the aqueous phase, the organic phase was dried over MgSO$_4$ and concentrated under reduced pressure. Recrystallisation from methanol and further purification via column chromatography (eluent: DCM) yielded a bright yellow solid as the product (Y$_9$ = 80 %).

^{1}H-NMR (400 MHz, CDCl$_3$): δ (ppm) = 10.52 (d, 1 H, C<u>H</u>O, J = 5.4 Hz), 7.97 (m, 1 H, C$_\phi$<u>H</u>), 7.7–7.46 (m, 2 H, C$_\phi$<u>H</u>), 7.40–7.24 (m, 10 H, C$_\phi$<u>H</u>), 6.98 (m, 1 H, C$_\phi$<u>H</u>).

^{13}C-NMR (100 MHz, CDCl$_3$): δ (ppm) = 191.6 (d, 1 C, <u>C</u>HO, ^{3}J = 19.6 Hz), 141.2 (d, 1 C, <u>C</u>$_\phi$CHO, ^{2}J = 26.6 Hz), 138.5 (d, 1 C, <u>C</u>$_\phi$P, J = 14.8 Hz), 136.14 (d, 2 C, <u>C</u>$_\phi$P, J = 9.9 Hz), 134.2–133.6 (6 C, <u>C</u>$_\phi$H), 130.6 (d, 1 C, <u>C</u>$_\phi$H, J = 3.9 Hz), 129.1–128.7 (7 C, <u>C</u>$_\phi$H).

^{31}P-NMR (80 MHz, CDCl$_3$): δ (ppm) = -13.39.

HRMS (ESI): m/z calculated for C$_{19}$H$_{16}$OP ([M+H]$^+$) = 291.09333 (100 %), 292.09668 (20.5 %), 293.10004 (2.0 %); m/z found = 291.09265 (100 %), 292.09589 (20.0 %), 293.09918 (1.6 %).

7.2.1.7 (R,R)-N,N'-2-(diphenylphosphino)benzyl-cyclohexane-1,2-diamine (11)

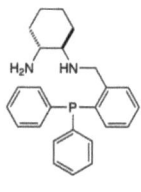

11

The reaction was performed in a different manner than described by Laue [88]. A vigorously stirred solution of diaminocyclohexane (10 mmol, 1.14 g) in DCM (100 mL) was cooled in an ice bath (temperature between 3 °C and 10 °C), the reaction being carried out under inert and anhydrous conditions. Over 5 h, a solution of compound **9** (9 mmol, 2.61 g) in DCM (50 mL) was added dropwise. After complete addition, the mixture was further stirred for 1.5 h keeping the temperature beneath 10 °C. DCM was then removed under reduced pressure, and the residue was dissolved in methanol (100 mL). The solution was again cooled in an ice bath, sodium borohydride (10 mmol, 0.38 g) was carefully added, and the mixture

7.2 Synthesis Procedures

was stirred for 20 h at room temperature. In order to quench the reaction, water (20 mL) was added. After extraction with DCM, the organic phase was dried over MgSO$_4$. Column chromatography (eluent: CHCl$_3$/MeOH/NH$_3$, 95/5/1) and concentration under reduced pressure yielded a clear, highly viscous, yellow oil (Y_{11} = 72 %).

^{1}H-NMR (400 MHz, CDCl$_3$): δ (ppm) = 7.62 (m, 1 H, C$_\phi$H), 7.51 (m, 1 H, C$_\phi$H), 7.4–7.2 (m, 10 H, C$_\phi$H), 7.15 (m, 1 H, C$_\phi$H), 6.88 (m, 1 H, C$_\phi$H), 4.10 (dd, 1 H, CH$_2$N, J = 1.5 Hz, J = 12.9 Hz), 3.91 (dd, 1 H, CH$_2$N, J = 2.0 Hz, J = 12.9 Hz), 2.27 (m, 1 H, CHN), 2.09–2.00 (m, 2 H, CHN/CH$_2$), 1.97 (br s, 3 H, NH/NH$_2$), 1.85 (m, 1 H, CH$_2$), 1.69–1.59 (m, 2 H, CH$_2$), 1.27–1.03 (m, 3 H, CH$_2$), 0.88 (m, 1 H, CH$_2$).

^{13}C-NMR (100 MHz, CDCl$_3$): δ (ppm) = 145.2 (d, C$_\phi$CH$_2$, J = 24.1 Hz), 137.0 (d, 1 C, C$_\phi$P, J = 8.6 Hz), 136.9 (d, 1 C, C$_\phi$P, J = 8.5 Hz), 135.7 (d, 1 C, C$_\phi$P, J = 13.3 Hz), 134.0–133.7 (5 C, C$_\phi$H), 129.3 (d, 1 C, C$_\phi$H, J = 5.4 Hz), 129.1–128.5 (7 C, C$_\phi$H), 127.2 (C$_\phi$H), 63.4 (CHNH), 55.4 (CHNH$_2$), 49.6 (d, 1 C, CH$_2$N, J = 21.6 Hz), 35.2 (CH$_2$), 31.36 (CH$_2$), 25.3 (CH$_2$), 25.2 (CH$_2$).

^{31}P-NMR (80 MHz, CDCl$_3$): δ (ppm) = -17.85.

HRMS (ESI): m/z calculated for C$_{25}$H$_{30}$N$_2$P ([M+H]$^+$) = 389.21411 (100 %), 390.21747 (27.0 %); m/z found = 389.21329 (100 %), 390.21631 (28.0 %).

7.2.1.8 (R,R)-N-[2-(Diphenylphosphino)benzyl-5-(hex-5-ynoic acid *tert*-butyl ester)]-N'-(2-diphenylphosphino)benzyl-cyclohexane-1,2-diamine (12)

A solution of compound **11** (2 mmol, 0.78 g) and compound **7** (2 mmol, 0.91 g) in 15 mL of DCM was stirred at room temperature under inert and anhydrous conditions for 16 h. DCM was then removed under reduced pressure, and the residue was dissolved in methanol (20 mL). While the flask was being cooled in an ice bath, sodium borohydride (10 mmol, 0.38 g) was added carefully. The mixture was stirred at room temperature, and the reaction was quenched through the addition of water (20 mL) after 20 h. After extraction with DCM, the organic phase was dried over MgSO$_4$. Column chromatography (eluent: CHCl$_3$/MeOH/NH$_3$, 95/5/1) and solvent removal gave a slightly yellow, foamy solid in 80 % yield.

^{1}H-NMR (400 MHz, CDCl$_3$): δ (ppm) = 7.55–7.10 (m, 25 H, C$_\phi$H), 6.82 (m, 1 H, C$_\phi$H), 6.75 (m, 1 H, C$_\phi$H), 4.04 (d, 1 H, J = 13.9 Hz, CH$_2$N), 3.86 (d, 1 H, CH$_2$N, J = 13.3 Hz), 3.94 (d, 1 H, CH$_2$N, J = 13.6

Hz) 3.78 (dd, 1 H, C<u>H</u>₂N, J = 1.3 Hz, J = 13.5 Hz), 2.42 (t, 2 H, C<u>H</u>₂COOtBu, J = 7.0 Hz), 2.37 (t, 2 H, C<u>H</u>₂C_{alkynyl}, J = 7.5 Hz), 2.21–2.10 (br m, 2 H, C<u>H</u>N), 1.98 (m, 2 H, C<u>H</u>₂), 1.84 (p, 2 H, CH₂C<u>H</u>₂CH₂, J = 7.3 Hz), 1.67–1.53 (m, 2 H, C<u>H</u>₂), 1.44 (s, 9 H, C<u>H</u>₃), 1.3–1.01 (m, 2 H, C<u>H</u>₂), 0.97–0.82 (br m, 2 H, C<u>H</u>₂).

¹³C-NMR (100 MHz, CDCl₃): δ (ppm) = 172.5 (<u>C</u>OOtBu), 137.0–136.5 (5 C, tert C_ϕ), 135.6–135.3 (3 C, tert C_ϕ), 134.0–133.8 (8 C, <u>C</u>_ϕH), 133.3 (d, 2 C, <u>C</u>_ϕH, J = 8.1 Hz), 131.9 (d, 1 C, <u>C</u>_ϕH, J = 4.9 Hz), 130.0 (<u>C</u>_ϕH), 129.04–128.45 (15 C, <u>C</u>_ϕH), 127.0 (<u>C</u>_ϕH), 124.5 (<u>C</u>_ϕC_{alkynyl}), 90.1 (<u>C</u>_{alkynyl}CH₂), 81.2 (<u>C</u>_{alkynyl}C_ϕ), 80.2 [<u>C</u>(CH₃)₃], 61.2 (<u>C</u>HN), 60.9 (<u>C</u>HN), 49.0 (d, 2 C, <u>C</u>H₂N, J = 22.3 Hz), 34.5 (<u>C</u>H₂), 31.3 (2 C, <u>C</u>H₂), 28.1 (3 C, <u>C</u>H₃), 25.0 (<u>C</u>H₂), 24.9 (<u>C</u>H₂), 24.2 (<u>C</u>H₂), 19.0 (<u>C</u>H₂).

³¹P-NMR (80 MHz, CDCl₃): δ (ppm) = -17.71.

HRMS (ESI): m/z calculated for C₅₄H₅₉N₂O₂P₂ ([M+H]⁺) = 829.40463 (100 %), 830.40798 (58.4 %), 831.41134 (16.7 %); m/z found = 829.40594 (100 %), 830.40881 (59.4 %), 831.41211 (16.5 %).

7.2.1.9 (R,R)-N-[2-(Diphenylphosphino)benzyl-5-(hex-5-ynoic acid)]-N'-(2-diphenylphosphino)benzyl-cyclohexane-1,2-diamine (13)

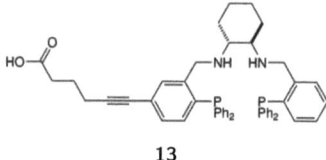

13

Under inert and anhydrous conditions, compound **12** (0.96 mmol, 0.80 g) was dissolved in DCM (6 mL), and trifluoroacetic acid (1.5 mL) was added. The solution was stirred at room temperature over night. After concentration under reduced pressure, the residue was dissolved in methanol (20 mL) and mixed with an aequeous solution of NaOH (0.1 M, 25 mL). Extraction with DCM, drying over MgSO₄ and concentration under reduced pressure yielded the product as a white foamed solid (Y₁₃ = 94 %).

¹H-NMR (400 MHz, CDCl₃): δ (ppm) = 7.70 (m, 1 H, C_ϕ<u>H</u>), 7.62 (m, 1 H, C_ϕ<u>H</u>), 7.37–7.10 (m, 23 H, C_ϕ<u>H</u>), 6.91 (m, 1 H, C_ϕ<u>H</u>), 6.75 (m, 1 H, C_ϕ<u>H</u>), 4.36 (d, 1 H, C<u>H</u>₂N, J = 13.3 Hz), 4.20 (dd, 1 H, C<u>H</u>₂N, J = 2.2 Hz, J = 13.3 Hz), 4.18 (dd, 1 H, C<u>H</u>₂N, J = 2.0 Hz, J = 13.0 Hz), 3.84 (d, 1 H, C<u>H</u>₂N, J = 13.0 Hz), 2.56 (m, 1 H, C<u>H</u>N), 2.52–2.40 (m, 5 H, C<u>H</u>N/C<u>H</u>₂COOtBu/C<u>H</u>₂C_{alkynyl}), 2.20–2.05 (m, 2 H, C<u>H</u>₂), 1.94–1.85 (m, 2 H, CH₂C<u>H</u>₂CH₂), 1.76–1.63 (m, 2 H, C<u>H</u>₂), 1.46–0.82 (m, 4 H, C<u>H</u>₂).

¹³C-NMR (100 MHz, CDCl₃): δ (ppm) = 176.1 (<u>C</u>OOH), 140.7 (d, <u>C</u>_ϕCH₂, J = 23.8 Hz), 138.5 (d, <u>C</u>_ϕCH₂, J = 24.6 Hz), 136.6 (d, <u>C</u>_ϕP, J = 13.8 Hz), 135.6–135.2 (5 C, tert C_ϕ), 134.1–133.4 (11 C, <u>C</u>_ϕH), 130.8–130.7 (3 C, <u>C</u>_ϕH), 130.0 (<u>C</u>_ϕH), 129.2–128.7 (12 C, <u>C</u>_ϕH), 125.3 (<u>C</u>_ϕC_{alkynyl}), 91.2

($\underline{C}_{alkynyl}$CH$_2$), 81.1 ($\underline{C}_{alkynyl}$C$_\phi$), 59.7 ($\underline{C}$HN), 59.2 ($\underline{C}$HN), 47.7 (d, 1 C, $\underline{C}$H$_2$N, J = 21.7 Hz), 46.7 (d, 1 C, $\underline{C}$H$_2$N, J = 22.5 Hz), 32.2 ($\underline{C}$H$_2$), 29.7 ($\underline{C}$H$_2$), 28.4 ($\underline{C}$H$_2$), 24.2 ($\underline{C}$H$_2$), 24.2 ($\underline{C}$H$_2$), 23.3 ($\underline{C}$H$_2$), 19.2 ($\underline{C}$H$_2$).

^{31}P-NMR (121 MHz, CDCl$_3$): δ (ppm) = -18.01/-18.62.

HRMS (ESI): m/z calculated for C$_{50}$H$_{51}$N$_2$O$_2$P$_2$ ([M+H]$^+$) = 773.34203 (100%), 774.34538 (54.1%), 775.34874 (14.3%); m/z found = 773.34100 (100%), 774.34412 (56.2%), 775.34741 (14.7%).

IR (ATR): ν (cm^{-1}) = 1717, 1665, 1588, 1434, 1386, 1196, 1132, 1026, 996, 830, 798, 743, 720, 696.

7.2.1.10 Dichloro{(R,R)-N-[2-(Diphenylphosphino)benzyl-5-(hex-5-ynoic acid)]-N'-(2-diphenylphosphino)benzyl-cyclohexane-1,2-diamine}ruthenium(II) (14)

14

Compound **13** (0.5 mmol, 0.38 g) and RuCl$_2$(DMSO)$_4$ (0.6 mmol, 0.29 g) were dissolved in toluene (40 mL). The solution was stirred under reflux conditions for 2 h and was then concentrated under reduced pressure. The residue, an orange-brown solid film, was purified via column chromatography (eluent: DCM/acetone, 10/2) and yielded the product, which appeared now yellow-brown (Y$_{14}$ = 65%).

^{1}H-NMR (400 MHz, CDCl$_3$): δ (ppm) = 7.34–7.18 (m, 10 H, C$_\phi$$\underline{H}$), 7.16–7.10 (m, 4 H, C$_\phi$$\underline{H}$), 7.08–6.94 (m, 13 H, C$_\phi$$\underline{H}$), 4.71 (m, 2 H, C$\underline{H}_2$N), 4.14–4.02 (m, 2 H, C$\underline{H}_2$N), 4.00–3.91 (br m, 2 H, C$\underline{H}$N), 3.04–2.97 (br m, 2 H, N$\underline{H}$), 2.82–2.75 (m, 2 H, C$\underline{H}_2$), 2.56–2.51 (t, 2 H, C$\underline{H}_2$COOtBu, J = 7.4 Hz), 2.51–2.47 (t, 2 H, C$\underline{H}_2$C$_{alkynyl}$, J = 6.9 Hz), 1.96–1.88 (m, 2 H, CH$_2$C$\underline{H}_2$CH$_2$), 1.88–1.83 (br m, 2 H, C$\underline{H}_2$), 1.32–1.14 (m, 4 H, C$\underline{H}_2$).

^{13}C-NMR (100 MHz, CDCl$_3$): δ (ppm) = 178.4 ($\underline{C}$OOH), 140.9 (d, 2 C, tert C$_\phi$, J = 14.5 Hz), 136.4–135.8 (6 C, tert C$_\phi$), 134.7–127.0 (C$_\phi$H), 124.9 ($\underline{C}_\phi$C$_{alkynyl}$), 90.4 ($\underline{C}_{alkynyl}$CH$_2$), 80.9 ($\underline{C}_{alkynyl}$C$_\phi$), 63.2 (2 C, $\underline{C}$HN), 52.4 (d, 1 C, $\underline{C}$H$_2$N, J = 4.8 Hz), 52.2 (d, 1 C, $\underline{C}$H$_2$N, J = 4.8 Hz), 32.7 ($\underline{C}$H$_2$), 30.9 ($\underline{C}$H$_2$), 30.9 ($\underline{C}$H$_2$), 24.9 ($\underline{C}$H$_2$), 24.9 ($\underline{C}$H$_2$), 23.6 ($\underline{C}$H$_2$), 18.9 ($\underline{C}$H$_2$).

^{31}P-NMR (121 MHz, CDCl$_3$): δ (ppm) = 40.88.

HRMS (ESI): m/z calculated for C$_{50}$H$_{50}$Cl$_2$N$_2$O$_2$P$_2$Ru ([M]$^+$) = 943.17749 (54.1%), 944.17626 (100%), 945.17961 (54.1%), 946.17331 (63.9%); m/z found = 943.17554 (62.0%), 944.17419 (100%), 945.17511

(66.4 %), 946.17328 (95.1 %).

IR (ATR): ν (cm^{-1}) = 1730, 1707, 1664, 1595, 1481, 1433, 1401, 1188, 1132, 1089, 1043, 961, 831, 743, 694.

7.2.1.11 Dichlorotetrakis(dimethyl sulphoxide)ruthenium(II)

RuCl$_2$(DMSO)$_4$ was prepared as described by Wilkinson et al. [198]. Ruthenium trichloride trihydrate was refluxed in dimethyl sulphoxide for 10 min. After cooling down to room temperature, the complex was precipitated from the dimethyl sulphoxide solution as an orange-yellow powder with acetone, and was purified by recrystallisation from hot dimethyl sulphoxide.

7.2.1.12 Di-µ-chloro-dichlorodihydridobis(cyclco-octa-I,5-diene)di-iridium(III)

[IrHCl$_2$(COD)]$_2$ was prepared as described by Robinson and Shaw [199]. Chloroiridic acid was dissolved in ethanol and heated to 95 °C for 30 min. Cyclo-octa-1,5-diene was added, and the mixture was stirred for further 2 h under reflux conditions. The flask was cooled down to about 0 °C, and the precipitate, a white-yellow powder, was filtered off and washed with ethanol and diethyl ether.

7.2.1.13 (R,R)-N,N'-bis[2-(Diphenylphosphino)benzyl]-cyclohexane-1,2-diamine (17)

Following a procedure used by Gao and co-workers [165], a mixture of (R,R)-1,2-diaminocyclohexane (3 mmol, 0.34 g), 2-(diphenylphosphino)benzaldehyde (6.0 mmol, 1.74 g), and Na$_2$SO$_4$ (18.0 mmol, 2.56 g) in DCM (15 mL) was stirred for 20 h under inert and anhydrous conditions. DCM was then removed under reduced pressure, and the residue was dissolved in methanol (15 mL). While the flask was cooled in an ice bath, sodium borohydride (30 mmol, 1.13 g) was added carefully. The mixture was stirred at room temperature, and the reaction was quenched by the addition of water (20 mL) after 20 h. After extraction with DCM, the organic phase was dried over MgSO$_4$. Column chromatography (eluent: DCM/acetone, 10/2) and solvent removal gave a slightly yellow, foamy solid in 80 % yield.

HRMS (ESI): m/z calculated for C$_{44}$H$_{45}$N$_2$P$_2$ ([M+H]$^+$) = 663.30525 (100 %), 664.30860 (47.6 %), 665.31196 (11.1 %); m/z found = 663.30432 (100 %), 664.30743 (47.9 %), 665.31091 (11.0 %).

7.2.1.14 Dichloro{(R,R)-N,N'-bis[2-(diphenylphosphino)benzyl]-cyclohexane-1,2-diamine}ruthenium(II) (18)

Similar to the procedure described by Noyori et al. [86], compound **18** was prepared by the reaction of **17** (0.48 g, 0.72 mmol) and RuCl$_2$(DMSO)$_4$ (0.53 g, 1.08 mmol) in toluene (20 mL) under reflux conditions for 6 h. After filtration, the solution was concentrated under reduced pressure. Silica gel chromatography of the crude product with DCM and acetone (2/1) afforded the product as yellow-orange crystals in 84 % yield. The same product (78 % yield) was obtained when the reaction was performed in 2-propanol instead of toluene under otherwise equal conditions.

HRMS (ESI): m/z calculated for $C_{44}H_{44}Cl_2N_2P_2Ru$ ([M]$^+$) = 833.14071 (54.1%), 834.13948 (100%), 835.14283 (47.6%); m/z found = 833.14075 (59.6%), 834.13971 (100%), 835.14075 (62.9%).

7.2.2 Synthesis of tethered Rh(III)-catalyst

A detailed description of the preparation of compound **20** is provided in the supporting information of reference [179]. Compound **20** was received from the group of Prof. Haag and was converted into the corresponding rhodium complex and the methylated complex, respectively.

7.2.2.1 Chloro(S,S)-⟨2-methyl-[2,3,4,5-tetramethylcyclopentadienyl]phenyl-{1,2-diphenyl-2-([4-amidophenyl sulfonyl]amido-hexanoic acid)-ethyl}amino⟩rhodium(III) (21)

21

Compound **20** (0.8 mmol, 674 mg) and RhCl$_3$·H$_2$O (0.96 mmol, 201 mg) were dissolved in THF (50 mL). After stirring for 20 h under reflux conditions, triethylamine (4 mmol, 554 µL) was added. The mixture was stirred for further 3 h and then concentrated under reduced pressure. Purification via column chromatography (eluent: dichloromethane/methanol, 9/1) yielded a red solid as the product (Y$_{21}$ = 50%).

HRMS (ESI): m/z calculated for $C_{42}H_{45}N_3O_5RhS$ ([M − Cl]$^+$) = 806.21295 (100%), 807.21630 (45.43%), 808.21966 (10.07%); m/z found = 806.21259 (100%), 807.21558 (47.82%), 808.21832 (12.27%).

IR (ATR): ν (cm^{-1}) = 1727, 1684, 1592, 1531, 1496, 1455, 1399, 1373, 1308, 1253, 1155, 1128, 1081, 1037, 1024, 935, 899, 835, 807, 750, 698.

7.2.2.2 Chloro(S,S)-⟨2-methyl-[2,3,4,5-tetramethylcyclopentadienyl]phenyl-{1,2-diphenyl-2-([4-amidophenyl sulfonyl]amido-hexanoic acid methyl ester)-ethyl}amino⟩rhodium(III) (23)

Compound **20** (0.25 mmol, 175 mg) and RhCl$_3$·H2O (0.25 mmol, 52 mg) were dissolved in methanol (20 mL). After stirring for 20 h under reflux conditions, triethylamine (70 µL, 0.5 mmol) was added. The

7 Experimental

23

mixture was stirred for further 3 h and then concentrated under reduced pressure. Purification via column chromatography (eluent: dichloromethane/methanol, 9/1) yielded a red solid as the product ($Y_{23} = 60\%$).

^{1}H-NMR (400 MHz, CDCl$_3$): δ (ppm) = 7.51 (dd, 1H, C$_\phi$H, J = 6.2Hz, J = 14.4Hz), 7.43 (d, 1 H, C$_\phi$H, J = 7.1 Hz), 7.34 (dd, 2 H, C$_\phi$H, J = 5.1 Hz, J = 11.3 Hz), 7.26–7.05 (m, 8 H, C$_\phi$H), 6.90 (d, 1H, C$_\phi$H, J = 7.4 Hz), 6.70 (m, 1 H, C$_\phi$H), 6.61 (t, 2 H, C$_\phi$H, J = 7.6 Hz), 6.45 (d, 2 H, C$_\phi$H, J = 7.5 Hz), 5.00 (br d, 1 H, CH$_2$NH, J = 12.5 Hz), 4.27 (d, 1 H, N$_{tosyl}$CH, J = 11.1 Hz), 4.22 (br d, 1 H, NHCH$_2$, J = 13.8 Hz), 3.66 (m, 1 H, NHCH$_2$), 3.56 (s, 3 H, OCH$_3$), 3.24 (t, 1 H, NHCH, J = 11.7 Hz), 2.40–2.25 (m, 4 H, CH$_2$CH$_2$CH$_2$CH$_2$), 2.04 (s, 3 H, CpCH$_3$), 1.92 (s, 3 H, CpCH$_3$), 1.80 (s, 3 H, CpCH$_3$), 1.67 (br s, 4 H, CH$_2$CH$_2$CH$_2$CH$_2$), 1.52 (s, 3 H, CpCH$_3$).

^{13}C-NMR (100 MHz, CDCl$_3$): δ (ppm) = 174.1 (NCO), 171.2 (COOCH$_3$), 139.9–135.3 (5 C, tert C$_\phi$), 126.7 (C$_\phi$), 131.5–126.6 (C$_\phi$H), 118.2 (C$_\phi$H), 106.4 (CCH$_3$), 99.4 (CCH$_3$), 97.1 (CCH$_3$), 88.3 (CCH$_3$), 80.7 (d, CCH$_3$), 75.8 (CHN$_{tosyl}$), 69.8 (CHNCH$_2$), 52.1 (CH$_2$N), 51.6 (COOCH$_3$), 37.0 (CH$_2$), 33.7 (CH$_2$), 24.8 (CH$_2$), 24.3 (CH$_2$), 10.6 (CH$_3$C), 10.5 (CH$_3$C), 10.2 (CH$_3$C), 8.1 (CH$_3$C).

HRMS (ESI): m/z calculated for C$_{43}$H$_{47}$N$_3$O$_5$RhS ([M − Cl]$^+$) = 820.22860 (100 %), 821.23195 (46.51 %), 822.23531 (10.56 %); m/z found = 820.22742 (100 %), 821.23041 (48.20 %), 822.23309 (11.13 %).

IR (ATR): ν (cm^{-1}) = 1735, 1693, 1592, 1529, 1497, 1455, 1401, 1354, 1309, 1255, 1157, 1130, 1090, 1025, 1004, 935, 912, 839, 807, 760, 731, 700.

7.3 Characterization of Supports and Coupling

7.3.1 Determination of Amine Loading

7.3.1.1 Cleavage of Protective Groups

Surface-functionalized support materials were received with Boc-protected amino groups. Cleavage was performed in a beaker. For example, six PCs (sinter chips, amine loading 1 μmol cm^{-2}, 2×2 cm^2 each) were added to 75 mL of a 6 N solution of HCl in 2-propanol. After stirring for 45 minutes, the PCs were accurately washed with methanol and dried under reduced pressure.

7.3.1.2 Qualitative Test

A solution of diisopropylethylamine (DIPEA, 10 %) in DMF and a 1 M solution of 2,4,6-trinitrobenzenesulfonic acid (TNBS) in DMF (about 2 mL in each case) were mixed in a beaker at room temperature. A small sample of the deprotected support material was added, and the mixture was left for 10 min. The support material was washed, and the amine loading was classified by the order of staining: intensely orange (high loading), yellow or slightly orange (moderate or low loading), colorless (no loading).

7.3.1.3 Quantification of Accessible Amine-Groups

Using the example of beads, a sample of about 50 mg of deprotected PBs were accurately weighed in an Eppendorf micro test tube and mixed with 0.25 mL of a 0.8 M solution of DIPEA in DMF. To this suspension, 35 μL of a fresh 0.63 M solution of Fmoc-β-Alanin-OPfp in N-methyl-2-pyrrolidone (NMP) were added, and the tube was vigorously shaken. After 20 min further 35 μL of the same solution were added, the tube was shaken again, and left for another 20 min. Then the tube was centrifugalized, and the solution was carefully decanted. Great importance was attached to the adherence to the following washing procedure: the particles-containing tube was filled with 0.8 mL of an organic solvent, vigorously shaken, centrifugalized, and the solvent was decanted. The washing was conducted five times using DMF (twice), then ethanol (once), and again DMF (twice). As cleavage reagent piperidine (0.8 mL of a 2.3 M solution in DMF) was added, the tube was shaken and left for 45 min. The solution was filled in a UV-transparant cuvette using a syringe filter to ensure that it was particle-free. The concentration of the cleaved fluorenyl derivative was calculated via the Lambert-Beer law ($\epsilon = 7800$ L mol^{-1} cm^{-1}) from the absorbtion determined at $\lambda = 301$ nm.

7.3.2 Immobilization and Determination of Catalyst Loading

7.3.2.1 General Procedure for Immobilization

The polymeric chips were added to a solution of DIPEA (145 μmol, 25 μL) in DMF (30 mL) in an atmosphere of argon. Separately compound **21** (32 μmol, 27 mg) and o-(benzotriazol-1-yl)-N,N,N',N'-tetramethyluronium tetrafluoroborate (TBTU; 36 μmol, 11.6 mg) were dissolved in DMF, and the solution was added to the PCs in DIPEA/DMF. After tentative shaking the flask for 20 h, the PCs were removed from the solution, washed with DMF and methanol, and dried under reduced pressure. Via ICP-based analysis of the rhodium content on the supports, a yield of up to 60 % was determined.

IR (ATR): ν (cm^{-1}) = 1651, 1590, 1535, 1494, 1453, 1401, 1371, 1307, 1252, 1214, 1148, 1125, 1099, 1034, 938, 895, 804, 759, 698.

7.3.2.2 General Procedure for ICP-OES Analysis

Pressurized digestion of catalyst-loaded support materials was performed by microwave treatment. A weighed sample of the supported catalyst (~100 mg) was mixed with nitric acid (65 % 5 mL) and hydrochloric acid

(37 % 1 mL). Within 5 min the sample was heated to 200 °C with a fixed microwave power of 200 W, and the temperature was hold for another 5 min. After digestion, the samples were filled up to 25 mL with water and subjected to ICP-OES analysis (usually two runs for each sample).

7.4 Catalysis Experiments

7.4.1 Homogeneous Catalysts

7.4.1.1 General Procedure for ATH Experiments Using the Ru-PNNP Catalyst

Complex **18** (10 µmol, 8.34 mg) was dissolved in 2-propanol (20 mL) under inert and anhydrous conditions. Acetophenone (5 mmol, 584 µL) was added, and the mixture was stirred at 45 °C. The reaction was started upon addition of a solution of potassium propan-2-olate (0.1 M, 200 µL).

7.4.1.2 General Procedure for ATH Experiments Using the Ir-PNNP Catalyst

Compound **17** (30 µmol, 20 mg) and [IrHCl$_2$(COD)]$_2$ (13.4 µmol, 10 mg) were dissolved in toluene (2 mL) and stirred for 1 h at 50 °C. A solution of sodium formate (1.9 mmol, 0.13 g) in water (20 mL) was added. The reaction was startet by addition of acetophenone (1.2 mmol, 0.14 mL).

7.4.2 Heterogenized Catalysts

7.4.2.1 General Procedure for ATH Experiments Using PC-supported Catalysts

The standard experiment was performed in a 25-mL flask using a magnetic stirrer with haeting. To a stirred solution of sodium formate (5 mmol, 340 mg) and acetophenone (1 mmol, 117 µL) in 10 mL of neat water at 40 °C, a catalyst-loaded PC (**22 b**, 2×2 cm^2) was added. Samples of 50 µL were taken from the reaction mixture, diluted with ethanol, and analyzed via GC-MS for determination of the conversion. The ee was determined from samples taken after extraction with ethyl acetate or DCM when the reaction was finished.

7.4.2.2 General Procedure for the Recycling of PC-supported Catalysts

After the reaction was finished, the catalyst chip was removed with forceps and placed in a beaker with 10 mL of methanol for about 5 minutes. Then, the chip was rinsed with methanol and dried under reduced pressure. When no subsequent experiment was performed, e. g., over night, the chip was stored in a refrigerator.

7.4.2.3 General Procedure for ATH Experiments Using PB-supported Catalysts

Experiments were carried out in a jacketed 100-mL flask equipped with a mechanical stirrer with stainless steel ringed propeller as well as an inside temperature gauge. In the standard experiment the temperature was adjusted to 40 °C, and a stirring speed of 1 200 rpm was applied. The reaction mixture consisted of 1.5 g of catalyst beads (**22 a**, rhodium content ≈ 0.105 mM) suspended in 100 mL of an aqueous solution of sodium formate (0.4 M) and acetophenone (0.04 M). The order of adding the components was as follows:

successively about 50 mL of dissolved sodium formate, the catalyst beads, and the remaining solution of sodium formate were added to the preheated reaction vessel. A couple of minutes of stirring was required to provide a homogeneous suspension at the designated temperature. The reaction was started by the addition of acetophenone. The reactor was kept closed except when samples (∼200 µL) were taken with a syringe. The samples were prepared for GC-MS analysis by filtering off the particles with a syringe filter.

Bibliography

[1] A. M. Rouhi, *Chem. Eng. News* **2004**, *82*, 47–62.

[2] R. A. Sheldon, I. Arends, U. Hanefeld, *Green Chemistry and Catalysis*, Wiley-VCH, Weinheim, **2007**, p. 1.

[3] B. Cornils, W. A. Herrmann, *J. Catal.* **2003**, *216*, 23–31.

[4] B. Cornils, W. A. Herrmann, I. T. Horváth, W. Leitner, S. Mecking, H. Olivier-Bourbigou, D. Vogt (Eds.), *Multiphase Homogeneous Catalysis, Vol. 1*, Wiley-VCH, Weinheim, **2005**, pp. 3–23.

[5] H.-U. Blaser, B. Pugin in *Handbook of Asymmetric Heterogeneous Catalysis*, K. Deng, Y. Uozumi (Eds.), Wiley-VCH, Weinheim, **2008**, pp. 413–437.

[6] E. L. Eliel, S. H. Wilen, *Organische Stereochemie*, Wiley-VCH, Weinheim, **1998**, pp. 533–583.

[7] H. Y. Aboul-Enein, I. W. Wainer, *The Impact of Stereochemistry on Drug Development and Use, Vol. 42*, John Wiley and Sons, New York, **1997**, pp. 10–17.

[8] FDA, *Development of New Stereoisomeric Drugs*, **5/1/1992**.

[9] J. Clayden, N. Greeves, S. Warren, P. Wothers, *Organic Chemistry*, Oxford University Press, **2001**, p. 386.

[10] Z. Wang, K. Ding, Y. Uozumi in *Handbook of Asymmetric Heterogeneous Catalysis*, K. Deng, Y. Uozumi (Eds.), Wiley-VCH, Weinheim, **2008**, pp. 1–24.

[11] E. L. Eliel, S. H. Wilen, *Organische Stereochemie*, Wiley-VCH, Weinheim, **1998**, pp. 185–289.

[12] W. J. Lough in *Chirality in Natural and Applied Scince*, W. J. Lough, I. W. Wainer (Eds.), Blackwell, **2002**, pp. 179–202.

[13] R. Noyori, H. Takaya, *Acc. Chem. Res.* **1990**, *23*, 345–350.

[14] H.-U. Blaser, B. Pugin, F. Spindler, *J. Mol. Cat. A: Chem.* **2005**, *231*, 1–20.

[15] A. Behr, *Angewandte homogene Katalyse*, Wiley-VCH, Weinheim, **2008**, pp. 34–47.

[16] J. Hagen, *Industrial Catalysis, 2nd ed.*, Wiley-VCH, Weinheim, **2006**, pp. 1–14.

[17] J. Weitkamp, R. Gläser in *Winnacker/Küchler: Chemische Technik, 5th ed.*, Vol. 1, R. Dittmeyer, W. Keim, G. Kreysa, A. Oberholz (Eds.), Wiley-VCH, Weinheim, **2004**, pp. 645–718.

[18] J. Hagen, *Industrial Catalysis, 2nd ed.*, Wiley-VCH, Weinheim, **2006**, p. 9.

[19] A. M. Palmer, A. Zanotti-Gerosa in *Asymmetric Catalysis on Industrial Scale: Challenges, Approaches, and Solutions, 2nd ed.*, H.-U. Blaser, H.-J. Federsel (Eds.), Wiley-VCH, Weinheim, **2010**, pp. 61–78.

[20] H.-U. Blaser, G. Hoge, B. Pugin, F. Spindler in *Green Catalysis - Homogeneous Catalysis*, Vol. 1 of *Handbook of Green Chemistry*, R. H. Crabtree (Ed.), Wiley-VCH, Weinheim, **2009**, pp. 153–203.

[21] A. Behr, *Angewandte homogene Katalyse*, Wiley-VCH, Weinheim, **2008**, p. 41.

[22] M. M. Green, H. A. Wittcoff, *Organic Chemistry Principles and Industrial Practice*, Wiley-VCH, Weinheim, **2003**, p. 249.

[23] W. Kersten, E.-M. Kern in *Betriebswirtschaftslehre für Chemiker*, G. Festel, A. Hassan, J. Leker, P. Bamelis (Eds.), Springer-Verlag, Berlin Heidelberg, **2001**, pp. 248–334.

[24] R. A. Sheldon, I. Arends, U. Hanefeld, *Green Chemistry and Catalysis*, Wiley-VCH, Weinheim, **2007**, p. 37.

[25] H.-U. Blaser, H.-J. Federsel (Eds.), *Asymmetric Catalysis on Industrial Scale, 2nd ed.*, Wiley-VCH, Weinheim, **2010**, pp. xxix–xxxvii.

[26] P. T. Anastas, J. C. Warner, *Green Chemistry: Theory and Practice*, Oxford University Press, New York, **1998**, p. 30.

[27] P. T. Anastas, J. B. Zimmerman, *Env. Sci. Tech.* **2003**, *37*, 94A–101A.

[28] A. Moores in *Green Catalysis - Homogeneous Catalysis*, Vol. 1 of *Handbook of Green Chemistry*, R. H. Crabtree (Ed.), Wiley-VCH, Weinheim, **2009**, pp. 1–15.

[29] R. A. Sheldon, *Chemistry and Industry* **1992**, 903–906.

[30] R. A. Sheldon, I. Arends, U. Hanefeld, *Green Chemistry and Catalysis*, Wiley-VCH, Weinheim, **2007**, pp. 2–5.

[31] C. Hedberg in *Modern Reduction Methods*, P. G. Andersson, I. J. Munslow (Eds.), Wiley-VCH, Weinheim, **2008**, pp. 109–134.

[32] R. Noyori, *Angew. Chem. Int. Ed.* **2002**, *41*, 2008–2022.

[33] A. Miyashita, A. Yasuda, H. Takaya, K. Toriumi, T. Ito, T. Souchi, R. Noyori, *J. Am. Chem. Soc.* **1980**, *102*, 7932–7934.

[34] R. Noyori, T. Ohkuma, *Angew. Chem. Int. Ed.* **2001**, *40*, 40–73.

[35] R. Noyori, T. Ohkuma, M. Kitamura, H. Takaya, N. Sayo, H. Kumobayashi, S. Akutagawa, *J. Am. Chem. Soc.* **1987**, *109*, 5856–5858.

[36] T. Ohkuma, H. Ooka, S. Hashiguchi, T. Ikariya, R. Noyori, *J. Am. Chem. Soc.* **1995**, *117*, 2675–2676.

[37] T. Ohkuma, M. Koizumi, K. Muñiz, G. Hilt, C. Kabuto, R. Noyori, *J. Am. Chem. Soc.* **2002**, *124*, 6508–6509.

[38] J. C. Fiaud, H. B. Kagan, *Bull. Soc. Chim. Fr.* **1969**, *8*, 2742–2743.

[39] C. R. Johnson, C. J. Stark, *Tetrahedron Lett.* **1979**, *20*, 4713–4716.

[40] A. Hirao, S. Itsuno, S. Nakahama, N. Yamazaki, *J. Chem. Soc., Chem. Commun.* **1981**, 315–317.

[41] E. J. Corey, C. J. Helal, *Angew. Chem. Int. Ed.* **1998**, *37*, 1986–2012.

[42] S. Itsuno in *Comprehensive Asymmetric Catalysis, Vol. 1*, E. N. Jacobsen, A. Pfaltz, H. Yamamoto (Eds.), Springer-Verlag, Berlin Heidelberg, **2000**, pp. 289–316.

[43] E. J. Corey, R. K. Bakshi, S. Shibata, *J. Am. Chem. Soc.* **1987**, *109*, 5551–5553.

[44] N. Arai, T. Ohkuma in *Modern Reduction Methods*, P. G. Andersson, I. J. Munslow (Eds.), Wiley-VCH, Weinheim, **2008**, pp. 159–181.

[45] H. Nishiyama in *Comprehensive Asymmetric Catalysis, Vol. 1*, E. N. Jacobsen, A. Pfaltz, H. Yamamoto (Eds.), Springer-Verlag, Berlin Heidelberg, **2000**, pp. 267–288.

[46] S. Rendler, M. Oestrich in *Modern Reduction Methods*, P. G. Andersson, I. J. Munslow (Eds.), Wiley-VHC, Weinheim, **2008**, pp. 183–208.

[47] A. S. Bommarius, B. R. Riebel, *Biocatalysis*, Wiley-VCH, Weinheim, **2004**, pp. 1–18.

[48] T. P. Yoon, E. N. Jacobsen, *Science* **2003**, *299*, 1691–1693.

[49] K. Nakamura, T. Matsuada in *Modern Reduction Methods*, P. G. Andersson, I. J. Munslow (Eds.), Wiley-VCH, Weinheim, **2008**, pp. 209–234.

[50] K. Nakamura, K. Takenaka, M. Fujii, Y. Ida, *Tetrahedron Lett.* **2002**, *43*, 3629–3631.

[51] J. C. Moore, D. J. Pollard, B. Kosjek, P. N. Devine, *Acc. Chem. Res.* **2007**, *40*, 1412–1419.

[52] B. Cornils, W. A. Herrmann, R. Schlögl, C.-H. Wong (Eds.), *Catalysis from A to Z, 2nd ed.*, Wiley-VCH, Weinheim, **2003**, p. 780.

[53] M. J. Palmer, M. Wills, *Tetrahedron Asymm.* **1999**, *10*, 2045–2061.

[54] S. Gladiali, E. Alberico, *Chem. Soc. Rev.* **2006**, *35*, 226–236.

[55] E. Knoevenagel, B. Bergdolt, *Chemische Berichte* **1903**, *36*, 2857.

[56] H. Meerwein, R. Schmidt, *Justus Liebigs Ann. Chem.* **1925**, *444*, 221–238.

[57] W. Ponndorf, *Angew. Chem.* **1926**, *39*, 138–143.

[58] A. Verly, *Bull. Soc. Chim. Fr.* **1925**, *37*, 537–542.

[59] R. V. Oppenauer, *Recl. Trav. Chim. Pays-Bas* **1937**, *56*, 137–144.

[60] W. von E. Doering, R. W. Young, *J. Am. Chem. Soc.* **1950**, *72*, 631.

[61] Y. Y. Haddad, H. B. Henbest, J. Husbands, T. R. B. Mitchell, *Proc. Chem. Soc.* **1964**, 361.

[62] M. J. Trocha-Grimshaw, H. B. Henbest, *Chem. Commun. (London)* **1967**, 544.

[63] Y. Sasson, J. Blum, *Tetrahedron Lett.* **1971**, *12*, 2167–2170.

[64] Y. Sasson, J. Blum, *J. Org. Chem.* **1975**, *40*, 1887–1896.

[65] R. L. Chowdhury, J.-E. Bäkvall, *J. Chem. Soc., Chem. Commun.* **1991**, 1063–1064.

[66] S. Gladiali, R. Taras in *Modern Reduction Methods*, P. G. Andersson, I. J. Munslow (Eds.), Wiley-VCH, Weinheim, **2008**, pp. 135–157.

[67] R. Noyori, M. Yamakawa, S. Hashiguchi, *J. Org. Chem.* **2001**, *66*, 7931–7944.

[68] K.-J. Haack, S. Hashiguchi, A. Fujii, T. Ikariya, R. Noyori, *Angew. Chem. Int. Ed.* **1997**, *36*, 285–288.

[69] M. Yamakawa, H. Ito, R. Noyori, *J. Am. Chem. Soc.* **2000**, *122*, 1466–1478.

[70] D. G. Blackmond, A. Armstrong, V. Coombe, A. Wells, *Angew. Chem. Int. Ed.* **2007**, *46*, 3798–3800.

[71] P. T. Anastas, M. M. Kirchhoff, *Acc. Chem. Res.* **2002**, *35*, 686–694.

[72] P. Dunn in *Green Solvents - Reactions in Water*, Vol. 5, C.-J. Li (Ed.), Wiley-VCH, Weinheim, **2010**, pp. 363–383.

[73] R. Breslow in *Green Solvents - Reactions in Water*, Vol. 5 of *Handbook of Green Chemistry*, C.-J. Li (Ed.), Wiley-VCH, Weinheim, **2010**, pp. 1–29.

[74] X. Wu, X. Li, W. Hems, F. King, J. Xiao, *Org. Biomol. Chem.* **2004**, *2*, 1818–1821.

[75] C. Bubert, J. Blacker, S. M. Brown, J. Crosby, S. Fitzjohn, J. P. Muxworthy, T. Thorpe, J. M. Williams, *Tetrahedron Lett.* **2001**, *42*, 4037–4039.

[76] T. Thorpe, J. Blacker, S. M. Brown, C. Bubert, J. Crosby, S. Fitzjohn, J. P. Muxworthy, J. M. Williams, *Tetrahedron Lett.* **2001**, *42*, 4041–4043.

[77] H. Y. Rhyoo, H.-J. Park, Y. K. Chung, *Chem. Commun.* **2001**, 2064–2065.

[78] X. Wu, J. Xiao in *Green Solvents - Reactions in Water*, Vol. 5 of *Handbook of Green Chemistry*, C.-J. Li (Ed.), Wiley-VCH, Weinheim, **2010**, pp. 105–149.

[79] X. Wu, J. Xiao, *Chem. Commun.* **2007**, 2449–2466.

[80] J. L. Namy, J. Souppe, J. Collin, H. B. Kagan, *J. Org. Chem.* **1984**, *49*, 2045–2049.

[81] D. A. Evans, S. G. Nelson, M. R. Gagne, A. R. Muci, *J. Am. Chem. Soc.* **1993**, *115*, 9800–9801.

[82] J. Yang, B. List, *Org. Lett.* **2006**, *8*, 5653–5655.

[83] S. Enthaler, G. Erre, M. K. Tse, K. Junge, M. Beller, *Tetrahedron Lett.* **2006**, *47*, 8095–8099.

[84] S. Zhou, S. Fleischer, K. Junge, S. Das, D. Addis, M. Beller, *Angew. Chem. Int. Ed.* **2010**, *49*, 8121–8125.

[85] C. Sui-Seng, F. Freutel, A. J. Lough, R. H. Morris, *Angew. Chem. Int. Ed.* **2008**, *47*, 940–943.

[86] J.-X. Gao, T. Ikariya, R. Noyori, *Organometallics* **1996**, *15*, 1087–1089.

[87] T. Ohkuma, R. Noyori in *Comprehensive Asymmetric Catalysis*, Vol. 1, E. N. Jacobsen, A. Pfaltz, H. Yamamoto (Eds.), Springer-Verlag, Berlin Heidelberg, **2000**, pp. 199–246.

[88] S. Laue, Dissertation, Forschungszentrum Jülich, **2002**.

[89] T. Ohkuma, M. Kitamura, R. Noyori in *New Frontiers in Asymmetric Catalysis*, K. Mikami, M. Lautens (Eds.), John Wiley and Sons, Hoboken, **2007**, pp. 1–32.

[90] S. Hashiguchi, A. Fujii, J. Takehara, T. Ikariya, R. Noyori, *J. Am. Chem. Soc.* **1995**, *117*, 7562–7563.

[91] K. Püntener, L. Schwink, P. Knochel, *Tetrahedron Lett.* **1996**, *37*, 8165–8168.

[92] J. Takehara, S. Hashiguchi, A. Fujii, S.-i. Inoue, T. Ikariya, R. Noyori, *Chem. Commun.* **1996**, 233–234.

[93] M. Palmer, T. Walsgrove, M. Wills, *J. Org. Chem.* **1997**, *62*, 5226–5228.

[94] D. G. I. Petra, P. C. J. Kamer, A. L. Spek, H. E. Schoemaker, P. W. N. M. van Leeuwen, *J. Org. Chem.* **2000**, *65*, 3010–3017.

[95] D. Matharu, D. Morris, A. Kawamoto, G. Clarkson, M. Wills, *Org. Lett.* **2005**, *7*, 5489–5491.

[96] X. Li, J. Blacker, I. Houson, X. Wu, J. Xiao, *Synlett* **2006**, 1155–1160.

[97] R. Noyori, S. Hashiguchi, *Acc. Chem. Res.* **1997**, *30*, 97–102.

[98] M. Yamakawa, I. Yamada, R. Noyori, *Angew. Chem. Int. Ed.* **2001**, *40*, 2818–2821.

[99] G. Zassinovic, G. Mestroni, S. Gladiali, *Chem. Rev.* **1992**, *92*, 1051–1069.

[100] J.-E. Bäckvall, *J. Organomet. Chem.* **2002**, *652*, 105–111.

[101] R. B. Woodward, N. L. Wendler, F. J. Brutschy, *J. Am. Chem. Soc.* **1945**, *67*, 1425–1429.

[102] E. C. Ashby, *Acc. Chem. Res.* **1988**, *21*, 414–421.

[103] C. F. de Graauw, J. A. Peters, H. van Bekkum, J. Huskens, *Synthesis* **1994**, 1007–1017.

[104] R. Cohen, C. R. Graves, S. T. Nguyen, J. M. L. Martin, M. A. Ratner, *J. Am. Chem. Soc.* **2004**, *126*, 14796–14803.

[105] G. Zassinovich, R. Bettella, G. Mestroni, N. Bresciani-Pahor, S. Geremia, L. Randaccio, *J. Organomet. Chem.* **1989**, *370*, 187–202.

[106] J. S. M. Samec, J.-E. Bäckvall, P. G. Andersson, P. Brandt, *Chem. Soc. Rev.* **2006**, *35*, 237–248.

[107] Y. R. S. Laxmi, J.-E. Bäckvall, *Chem. Commun.* **2000**, 611–612.

[108] D. Alonso, P. Brandt, S. Nordin, P. Andersson, *J. Am. Chem. Soc.* **1999**, *121*, 9580–9588.

[109] D. G. I. Petra, J. N. H. Reek, J.-W. Handgraaf, E. J. Meijer, P. Dierkes, P. C. J. Kamer, J. Brussee, H. E. Schoemaker, P. W. N. M. van Leeuwen, *Chem. Eur. J.* **2000**, *6*, 2818–2829.

[110] O. Pàmies, J.-E. Bäckvall, *Chem. Eur. J.* **2001**, *7*, 5052–5058.

[111] J. Wettergren, E. Buitrago, P. Ryberg, H. Adolfsson, *Chem. Eur. J.* **2009**, *15*, 5709–5718.

[112] X. F. Wu, J. K. Liu, D. Di Tommaso, J. A. Iggo, C. R. A. Catlow, J. Bacsa, J. L. Xiao, *Chem. Eur. J.* **2008**, *14*, 7699–7715.

[113] R. G. Wilkins, *Kinetics and Mechanism of Reactions of Transition Metal Complexes*, 2nd ed., VCH, Weinheim, **1991**, pp. 88–89.

[114] J. Blacker, J. Martin in *Asymmetric Catalysis on Industrial Scale*, H.-U. Blaser, E. Schmidt (Eds.), Wiley-VCH, Weinheim, **2004**, pp. 201–216.

[115] T. Ikariya, A. J. Blacker, *Acc. Chem. Res.* **2007**, *40*, 1300–1308.

[116] A. J. Blacker, P. Thompson in *Asymmetric Catalysis on Industrial Scale: Challenges, Approaches, and Solutions, 2nd ed.*, H.-U. Blaser, H.-J. Federsel (Eds.), Wiley-VCH, Weinheim, **2010**, pp. 265–289.

[117] K. B. Hansen, J. R. Chilenski, R. Desmond, P. N. Devine, E. J. J. Grabowski, R. Heid, M. Kubryk, D. J. Mathre, R. Varsolona, *Tetrahedron Asymm.* **2003**, *14*, 3581–3587.

[118] H. Yamashita, T. Ohtani, S. Morita, K. Otsubo, K. Kan, J. Matsubara, K. Kitano, Y. Kawano, M. Uchida, F. Tabusa, *Heterocycles* **2002**, *56*, 123–128.

[119] *Strem Chemicals, Inc.*, **1/20/2011**. http://www.strem.com/uploads/technical_notes/96-7650tech.pdf.

[120] *Takasago*, 1/20/2011. http://www.takasago.com/finechemicals/ligands_catalysts/501.pdf.

[121] *Johnson Matthey*, 1/20/2011. http://www.jmcatalysts.com/pharma/site.asp?siteid=540.

[122] D. E. de Vos, I. F. J. Vankelecom, P. A. Jacobs (Eds.), *Chiral Catalyst Immobilization and Recycling*, Vol. 1, Wiley-VCH, Weinheim, 2000, p. 320.

[123] B. Pugin, H.-U. Blaser in *Comprehensive Asymmetric Catalysis*, Vol. 3, E. N. Jacobsen, A. Pfaltz, H. Yamamoto (Eds.), Spinger-Verlag, Berlin Heidelberg, 2000, pp. 1367–1375.

[124] G. Oehme in *Comprehensive Asymmetric Catalysis*, Vol. 3, E. N. Jacobsen, A. Pfaltz, H. Yamamoto (Eds.), Spinger-Verlag, Berlin Heidelberg, 2000, pp. 1377–1386.

[125] A. Kirschning (Ed.), *Immobilized Catalysts: Solid Phases, Immobilization and Applications*, Vol. 242, Springer-Verlag, Berlin Heidelberg, 2004.

[126] M. Heitbaum, F. Glorius, I. Escher, *Angew. Chem. Int. Ed.* 2006, *45*, 4732–4762.

[127] K. Ding, Y. Uozumi (Eds.), *Handbook of Asymmetric Heterogeneous Catalysis*, Wiley-VCH, Weinheim, 2008.

[128] N. End, K.-U. Schöning in *Immobilized Catalysts*, of *Topics in Current Chemistry*, A. Kirschning (Ed.), Springer-Verlag, Berlin Heidelberg, 2004, pp. 241–271.

[129] D. J. Bayston, C. B. Travers, M. E. C. Polywka, *Tetrahedron Asymm.* 1998, *9*, 2015–2018.

[130] N. A. Cortez, G. Aguirre, M. Parra-Hake, R. Somanathan, *Tetrahedron Lett.* 2009, *50*, 2228–2231.

[131] A. J. Sandee, D. G. I. Petra, J. N. H. Reek, P. C. J. Kamer, P. W. N. M. v. Leeuwen, *Chem. Eur. J.* 2001, *7*, 1202–1208.

[132] P. N. Liu, P. M. Gu, F. Wang, Y. Q. Tu, *Org. Lett.* 2004, *6*, 169–172.

[133] P. N. Liu, J. G. Deng, Y. Q. Tu, S. H. Wang, *Chem. Commun.* 2004, 2070–2071.

[134] P.-N. Liu, P.-M. Gu, J.-G. Deng, Y.-Q. Tu, Y.-P. Ma, *Eur. J. Org. Chem.* 2005, *15*, 3221–3227.

[135] C.-F. Nie, J.-S. Suo, *Chin. J. Chem.* 2005, *23*, 315–320.

[136] X. H. Huang, J. Y. Ying, *Chem. Commun.* 2007, 1825–1827.

[137] S. Parambadath, A. Singh, *Catalysis Today* 2009, *141*, 161–167.

[138] J. Li, Y. Zhang, D. Han, Q. Gao, C. Li, *J. Mol. Catal. A: Chem.* 2009, *298*, 31–35.

[139] F. Michalek, A. Lagunas, C. Jimeno, M. A. Pericàs, *J. Mater. Chem.* 2008, *18*, 4692–4697.

[140] S. Laue, L. Greiner, J. Woltinger, A. Liese, *Adv. Synth. Catal.* 2001, *343*, 711–720.

[141] M. Aglietto, E. Chiellini, S. D'Antone, G. Ruggeri, R. Solaro, *Pure Appl. Chem.* **1988**, *60*, 415–430.

[142] P. Gamez, B. Dunjic, F. Fache, M. Lemaire, *J. Chem. Soc., Chem. Commun.* **1994**, 1417–1418.

[143] P. Gamez, B. Dunjic, C. Pinel, M. Lemaire, *Tetrahedron Lett.* **1995**, *36*, 8779.

[144] F. Locatelli, P. Gamez, M. Lemaire, *J. Mol. Catal. A: Chem.* **1998**, *135*, 89–98.

[145] K. Polborn, K. Severin, *Chem. Commun.* **1999**, 2481–2482.

[146] K. Polborn, K. Severin, *Chem. Eur. J.* **2000**, *6*, 4604–4611.

[147] K. Polborn, K. Severin, *Eur. J. Inorg. Chem.* **2000**, *2000*, 1687–1692.

[148] M. Tada, Y. Iwasawa, *J. Mol. Catal. A: Chem.* **2003**, *199*, 115–137.

[149] R. ter Halle, E. Schulz, M. Lemaire, *Synlett* **1997**, 1257–1258.

[150] C. Saluzzo, R. ter Halle, F. Touchard, F. Fache, E. Schulz, M. Lemaire, *J. Organomet. Chem.* **2000**, *603*, 30–39.

[151] Y. Arakawa, N. Haraguchi, S. Itsuno, *Tetrahedron Lett.* **2006**, *47*, 3239–3243.

[152] Y. Arakawa, A. Chiba, N. Haraguchi, S. Itsuno, *Adv. Synth. Catal.* **2008**, *350*, 2295–2304.

[153] N. Haraguchi, K. Tsuru, Y. Arakawa, S. Itsuno, *Org. Biomol. Chem.* **2009**, *7*, 69–75.

[154] H.-F. Zhou, Q.-H. Fan, Y.-Y. Huang, L. Wu, Y.-M. He, W.-J. Tang, L.-Q. Gu, A. S. Chan, *J. Mol. Catal. A: Chem.* **2007**, *275*, 47–53.

[155] H. Y. Rhyoo, H.-J. Park, W. H. Suh, Y. K. Chung, *Tetrahedron Lett.* **2002**, *43*, 269.

[156] F. Wang, H. Liu, L. Cun, J. Zhu, J. Deng, Y. Jiang, *J. Org. Chem.* **2005**, *70*, 9424–9429.

[157] H. Yang, J. Li, J. Yang, Z. Liu, Q. Yang, C. Li, *Chem. Commun.* **2007**, 1086–1088.

[158] S. Bai, H. Yang, P. Wang, J. Gao, B. Li, Q. Yang, C. Li, *Chem. Commun.* **2010**, *46*, 8145–8147.

[159] *PolyAn GmbH*, **1/22/2011**. http://www.poly-an.de.

[160] F. Koc, F. Michalek, L. Rumi, W. Bannwarth, R. Haag, *Synthesis* **2005**, 3362–3372.

[161] U. Schedler, *unpublished results*.

[162] W.-K. Wong, J.-X. Gao, Z.-Y. Zhou, T. C. W. Mak, *Polyhedron* **1992**, *11*, 2965–2966.

[163] W.-K. Wong, J.-X. Gao, Z.-Y. Zhou, T. C. Mak, *Polyhedron* **1993**, *12*, 1415–1417.

[164] J.-X. Gao, H.-L. Wan, W.-K. Wong, M.-C. Tse, W.-T. Wong, *Polyhedron* **1996**, *15*, 1241–1251.

[165] J.-X. Gao, X.-D. Yi, P.-P. Xu, C.-L. Tang, H.-L. Wan, T. Ikariya, *J. Organomet. Chem.* **1999**, *592*, 290–295.

[166] J.-X. Gao, H. Zhang, X.-D. Yi, P.-P. Xu, C.-L. Tang, H.-L. Wan, K.-R. Tsai, T. Ikariya, *Chirality* **2000**, *12*, 383–388.

[167] Z.-R. Dong, Y.-Y. Li, J.-S. Chen, B.-Z. Li, Y. Xing, J.-X. Gao, *Org. Lett.* **2005**, *7*, 1043–1045.

[168] Y. Xing, J. S. Chen, Z.-R. Dong, Y.-Y. Li, J.-X. Gao, *Tetrahedron Lett.* **2006**, *47*, 4501–4503.

[169] R. M. Stoop, A. Mezzetti, *Green Chemistry* **1999**, *1*, 39–41.

[170] R. M. Stoop, S. Bachmann, M. Valentini, A. Mezzetti, *Organometallics* **2000**, *19*, 4117–4126.

[171] W.-K. Wong, X.-P. Chen, T.-W. Chik, W.-Y. Wong, J.-P. Guo, F.-W. Lee, *Eur. J. Inorg. Chem.* **2003**, *2003*, 3539–3546.

[172] J.-X. Gao, X. D. Yi, C.-L. Tang, P.-P. Xu, H.-L. Wan, *Polym. Adv. Technol.* **2001**, *12*, 716–719.

[173] L. Greiner, S. Laue, A. Liese, C. Wandrey, *Chem. Eur. J.* **2006**, *12*, 1818–1823.

[174] I. Klement, P. Knochel, K. Chau, G. Cahiez, *Tetrahedron Lett.* **1994**, *35*, 1177–1180.

[175] S.-Y. Han, Y.-A. Kim, *Tetrahedron* **2004**, *60*, 2447–2467.

[176] X. F. Wu, D. Vinci, T. Ikariya, J. L. Xiao, *Chem. Commun.* **2005**, 4447–4449.

[177] D. S. Matharu, D. J. Morris, G. J. Clarkson, M. Wills, *Chem. Commun.* **2006**, 3232–3234.

[178] D. Morris, A. Hayes, M. Wills, *J. Org. Chem.* **2006**, *71*, 7035–7044.

[179] J. Dimroth, J. Keilitz, U. Schedler, R. Schomäcker, R. Haag, *Adv. Synth. Catal.* **2010**, *352*, 2497–2506.

[180] Y.-C. Chen, T.-F. Wu, J.-G. Deng, H. Liu, X. Cui, J. Zhu, Y.-Z. Jiang, M. C. K. Choi, A. S. C. Chan, *J. Org. Chem.* **2002**, *67*, 5301–5306.

[181] T. Koike, T. Ikariya, *Adv. Synth. Catal.* **2004**, *346*, 37–41.

[182] S. Richards, M. Ropic, D. Blackmond, A. Walmsley, *Anal. Chim. Acta* **2004**, *519*, 1–9.

[183] A. J. Blacker, S. B. Duckett, J. Grace, R. N. Perutz, A. C. Whitwood, *Organometallics* **2009**, *28*, 1435–1446.

[184] X. Wu, X. Li, F. King, J. Xiao, *Angew. Chem. Int. Ed.* **2005**, *44*, 3407–3411.

[185] J. Canivet, G. Labat, H. Stoeckli-Evans, G. Süss-Fink, *Eur. J. Inorg. Chem.* **2005**, *2005*, 4493–4500.

[186] X. Wu, X. Li, A. Zanotti-Gerosa, A. Pettman, J. Liu, A. Mills, J. Xiao, *Chem. Eur. J.* **2008**, *14*, 2209–2222.

[187] X. F. Wu, J. K. Liu, X. H. Li, A. Zanotti-Gerosa, F. Hancock, D. Vinci, J. W. Ruan, J. L. Xiao, *Angew. Chem. Int. Ed.* **2006**, *45*, 6718–6722.

[188] P. Wessig, *unpublished results*.

[189] K. A. Connors, *Chemical Kinetics*, Wiley-VCH, **1990**, p. 220.

[190] J. Canivet, G. Süss-Fink, *Green Chemistry* **2007**, *9*, 391–397.

[191] L. Arnaut, H. Burrows, S. Formosinho, *Chemical Kinetics: From Molecular Structure to Chemical Reactivity*, Elsevier, Amsterdam, **2007**, p. 244.

[192] L. A. Campbell, T. Kodadek, *J. Mol. Catal. A: Chem.* **1996**, *113*, 293–310.

[193] M. Watanabe, K. Murata, T. Ikariya, *J. Am. Chem. Soc.* **2003**, *125*, 7508–7509.

[194] M. Watanabe, A. Ikagawa, H. Wang, K. Murata, T. Ikariya, *J. Am. Chem. Soc.* **2004**, *126*, 11148–11149.

[195] G. Bartoli, M. Bosco, A. Carlone, R. Dalpozzo, E. Marcantoni, P. Melchiorre, L. Sambria, *Synthesis* **2007**, 3489–3496.

[196] B. Neises, W. Steglich, *Angew. Chem. Int. Ed.* **1978**, *17*, 522–524.

[197] D. B. Dess, J. C. Martin, *J. Org. Chem.* **1983**, *48*, 4155–4156.

[198] I. P. Evans, A. Spencer, G. Wilkinson, *J. Chem. Soc., Dalton Trans.* **1973**, 204–209.

[199] S. D. Robinson, B. L. Shaw, *J. Chem. Soc.* **1965**, 4997–5001.

Appendix

Publications out of this Work

Conference Contributions

- J. Dimroth, J. Keilitz, R. Haag, P. Wessig, R. Schomäcker, *Asymmetric Transfer Hydrogenation with Heterogenised Catalysts*, 8th European Congress of Chemical Engineering, 2011, Berlin; poster presentation.

- J. Dimroth, M. Schwarze, J. Keilitz, R. Haag, R. Schomäcker, *Heterogeneous asymmetric transfer hydrogenation*, 44. Jahrestreffen Deutscher Katalytiker mit Jahrestreffen Reaktionstechnik, 2011, Weimar; poster presentation.

- M. Schwarze, J. Keilitz, J. Dimroth, S. Nowag, U. Schedler, R. Haag, R. Schomäcker, Recyclable Metal Catalysts for Hydrogenation Reactions, Joint International Symposium *Activation of Small Molecules – Gas Phase Clusters, Molecular Catalysts, Enzymes and Solid Materials*, 2011, Erkner; poster presentation.

- J. Dimroth, J. Keilitz, U. Schedler, R. Haag, R. Schomäcker, *Oberflächenfunktionalisierte Polymere als Träger für chirale Katalysatoren - Anwendung in der asymmetrischen Transferhydrierung*, ProcessNet-Jahrestagung, 2010, Aachen; talk.

- J. Dimroth, J. Keilitz, R. Haag, R. Schomäcker, *Heterogenisation of a Rh(III)-catalyst and application to asymmetric transfer hydrogenation*, 43. Jahrestreffen Deutscher Katalytiker, 2010, Weimar; poster presentation.

- M. Schwarze, J. S. Milano-Brusco, H. Nowothnick, A. Rost, K. Seifert, S. Jost, J. Dimroth, R. Schomäcker, *Katalyse in Tensidsystemen*, 43. Jahrestreffen Deutscher Katalytiker, 2010, Weimar; poster presentation.

- J. Dimroth, J. Keilitz, R. Haag, R. Schomäcker, *An easily recyclable heterogenised Rh-catalyst for the asymmetric transfer hydrogenation of prochiral ketones*, ISHHC XIV, 2009, Stockholm; poster presentation.

Publications in Journals

- J. Dimroth, U. Schedler, J. Keilitz, R. Haag, R. Schomäcker, *New Polymer-Supported Catalysts for the Asymmetric Transfer Hydrogenation of Acetophenone in Water – Kinetic and Mechanistic Investigations*, Advanced Synthesis & Catalysis **2011**, *353*, 8, 1335–1344.

- J. Dimroth, J. Keilitz, U. Schedler, R. Schomäcker, R. Haag, *Immobilization of a Modified Tethered Rhodium(III)-p-Toluenesulfonyl-1,2-diphenylethylenediamine Catalyst on Soluble and Solid Polymeric Supports and Successful Application to Asymmetric Transfer Hydrogenation of Ketones*, Advanced Synthesis & Catalysis **2010**, *352*, 14-15, 2497–2506; paper featured in Synfacts **2011**, *1*, 112.

Patent Applications

- J. Keilitz, R. Haag, U. Schedler, J. Dimroth, *Immobilized Rhodium(III), Ruthenium(II), or Iridium(III) Catalysts for Asymmetric Hydrogenation Reactions*, WO/2011/026682.

- J. Dimroth, U. Schedler, *Immobilisierte Rhodium(III)-Katalysatoren für asymmetrische Hydrierreaktionen*, DE 10 2007 041 965 A1.

VDM Verlagsservicegesellschaft mbH

Die VDM Verlagsservicegesellschaft sucht für wissenschaftliche Verlage abgeschlossene und herausragende

Dissertationen, Habilitationen, Diplomarbeiten, Master Theses, Magisterarbeiten usw.

für die kostenlose Publikation als Fachbuch.

Sie verfügen über eine Arbeit, die hohen inhaltlichen und formalen Ansprüchen genügt, und haben Interesse an einer honorarvergüteten Publikation?

Dann senden Sie bitte erste Informationen über sich und Ihre Arbeit per Email an *info@vdm-vsg.de*.

Sie erhalten kurzfristig unser Feedback!

VDM Verlagsservicegesellschaft mbH
Dudweiler Landstr. 99
D - 66123 Saarbrücken
www.vdm-vsg.de

Telefon +49 681 3720 174
Fax   +49 681 3720 1749

Die VDM Verlagsservicegesellschaft mbH vertritt

Printed by Books on Demand GmbH, Norderstedt / Germany